Recreational Problems in

GEOMETRIC DISSECTIONS and How to Solve Them

HARRY
LINDGREN

Revised and enlarged by GREG FREDERICKSON

DOVER PUBLICATIONS, INC.
NEW YORK

FOR EVE AND JUDY

Published in Canada by General Publishing Company, Ltd., 30 Lesmill Road, Don Mills, Toronto, Ontario.
Published in the United Kingdom by Constable and Company, Ltd., 10 Orange Street, London WC 2.

This Dover edition, first published in 1972, is a revised and enlarged republication of the work originally published by the D. Van Nostrand Company, Inc., in 1964 under the title *Geometric Dissections*. The present edition contains a new preface, a new Appendix H and additions and corrections in Appendices E and F, all by Greg Frederickson.

International Standard Book Number: 0-486-22878-9
Library of Congress Catalog Card Number: 72-85502

Manufactured in the United States of America
Dover Publications, Inc.
180 Varick Street
New York, N. Y. 10014

PREFACE TO THE DOVER EDITION

Several years ago I picked up the first edition of this book. It immediately fascinated me. Not that I really understood how the dissections were figured out—I guess I was just enticed by the diagrams. Off and on I would pull the book down off my bookshelf and spend an hour marveling at the beauty of the dissections. I especially liked the chapter on star dissections, and I wondered how anyone could ever figure out something so beautiful. I never thought I could.

But then, perhaps dissections always happen by accident. In the early morning hours of May 30, 1971, I was fiddling around with Harry Lindgren's dissection of {8/2} into {4} (Fig. 19.6). All of a sudden I realized I had reduced by one the number of pieces needed for the dissection. A wild "eureka!" exploded: I had my first dissection record! Many more have come since then. Always they have been unexpected—coming before work, late at night, during meals, or at other times when something else was planned.

And after every couple of dissections I would hurry off to get them copied and then mail them halfway around the world to Harry Lindgren in Australia. It must have seemed strange to him: to have written a book eight years ago which was sliding comfortably onto the recreational math book lists; then, suddenly, a profusion of letters from America, dashed off by an excited young man who seemed afraid he might die before the latest dissection was in someone else's hands. Harry Lindgren, who had turned to other interests after the publication of his book, patiently answered all these letters, with understanding and encouragement. For he knew the power that such a subject could exercise on one.

Since I was the one who rendered parts of the work out of date, Harry Lindgren assigned me the responsibility of revising his book. This I have done by adding an Appendix H and correcting and adding entries in Appendices E and F. This method of revision was chosen over correcting the main part of the text, so that the book could be republished quickly and inexpensively. Moreover, the out-of-date dissections remaining in the text provide an unexpected advantage; by comparing the new dissections with those in the text, the reader can see how the dissections were improved upon and how gaps in the text were filled in. For the ambitious reader this leaves a challenge: gaps still remain and certainly not all the dissections are minimal. Who will be the one to write the *next* appendix?

I would like to thank Joseph S. Madachy, editor of the *Journal of Recreational Mathematics,* for encouraging me to write the articles from which much of Appendix H comes. Thanks also go to Martin Gardner, who was instrumental in getting this book reprinted. Finally, a special word of thanks to Harry Lindgren, who through his book introduced me to a fascinating pastime.

GREG FREDERICKSON

Baltimore, Maryland
May, 1972

PREFACE TO THE FIRST EDITION

Les matières de Géométrie sont si sérieuses d'elles-mêmes, qu'il est avantageux qu'il s'offre quelque occasion pour les rendre un peu divertissantes.

PASCAL

Professor Klotzkopf is mythical, but never mind. He achieved renown through his profound researches into integral equations of Volterra's first kind. These are difficult, but the Professor's hobby afforded relaxation and enabled him to return, with mind refreshed, to his lifework. His hobby was integral equations of Volterra's second kind.

You who are reading this would rather relax less strenuously, I am sure, but you are too brainy for Bingo. Then try your hand at dissections. Guided by this book, you should find in them a pastime that is neither too exacting nor too trivial. You may not be content merely to read about dissections, but would like to do some. If so, Appendix B will remove most of the drudgery involved; thanks to its help, you can plunge instead straight *in medias res.*

Not every kind of dissection falls within the scope of this book. Its subject is the problem of finding a dissection of one given plane figure into another in the fewest possible pieces—a *minimal* dissection—and it is quite strictly adhered to. Thus, the dissection of a square into unequal squares or into acute-angled triangles is not discussed.

The foregoing is not all I have to say in general about dissections. There is more in the Postscript.

If you like my book, thank Joseph S. Madachy, editor of *Recreational Mathematics Magazine,* for prodding me into writing it. A word of thanks is due also to Martin Gardner, whose article in the November 1961 *Scientific American* paved the way, and whose books on Sam Loyd's puzzles were indispensable in preparing this one. Others to whom I am indebted include Dr. C. Dudley Langford and Thomas H. O'Beirne, for information about Freese's dissections and solid dissections respectively.

HARRY LINDGREN

Patent Office, Canberra
December, 1963

CONTENTS

1

INTRODUCTORY

It was Hilbert (or was it Bolyai and Gerwien?) who first proved that any rectilinear plane figure can be dissected into any other of the same area by cutting it into a finite number of pieces. (It is natural to ask, did we need a Hilbert to do that?) In the proof no account is taken of the number of pieces—one wants only to show that it is finite. But the main interest of dissections as a recreation is to find how to dissect one figure into another in the least number of pieces. In a few cases (very, very few) it could perhaps be rigorously proved that the minimum number has been attained, and in a few more one can feel morally certain; in all the rest it is possible that you may find a dissection that is better than those already known, or may even find a new *kind* of dissection. The subject is nowhere near exhaustion. In this respect it compares favorably with many recreations of the algebraic kind; magic squares, for instance, have been worked almost to death.

The dissections one sees now and then in books and magazines may seem of infinite variety, but most of them are elaborations of only a few basic dissections, usually of one parallelogram into another. It is also heartening to know that there are more or less methodical procedures for finding a dissection. This book is about them.

The figures to be dissected will nearly always be referred to as *polygons,* where we are concerned for the moment with figures in general and not with a particular one. (We don't usually think of a Greek cross, say, as a polygon, but after all it is a dodecagon.) Since the main interest is in regular polygons, the naming of a particular polygon indicates the regular one, except where it clearly does not; e.g. *pentagon* will usually mean *regular pentagon.* And all polygons are rectilinear except in Chapter 23.

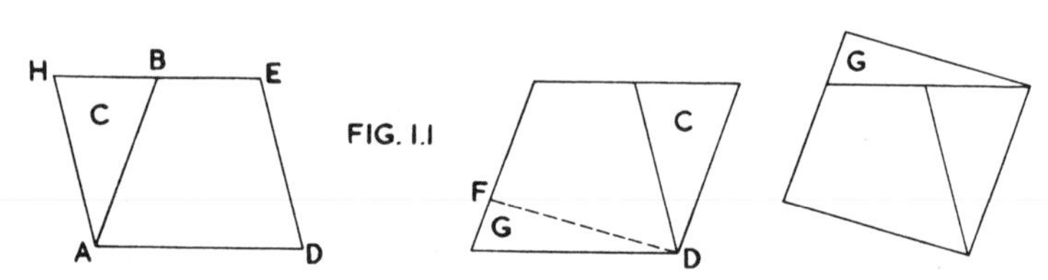

1.1. What I call an *S-dissection,* for a reason to appear in Chapter 2, is shown in Fig. 1.1. It is one of the ways of changing one parallelogram into another. We make a first cut AB equal to one side of the other parallelogram, and transfer piece C to the opposite side DE. The parallelogram thus modified has the desired base of length AB and the desired height perpendicular to AB, but the angles may be wrong. So we make a second cut DF (equal to the other side of the other parallelogram), and transfer piece G to the opposite, upper side. The parallelogram thus obtained may well have no side or angle equal to any side or angle in the original one.

As in Fig. 1.1, added cuts will be shown by broken lines (e.g. DF), where it is helpful to distinguish them from existing cuts.

The dimensions of the two parallelograms may be such that a three-piece dissection cannot be obtained. Thus, dissecting a 5×1 rectangle into a square requires four pieces. We make a first cut through J in Fig. 1.2 whose perpendicular distance

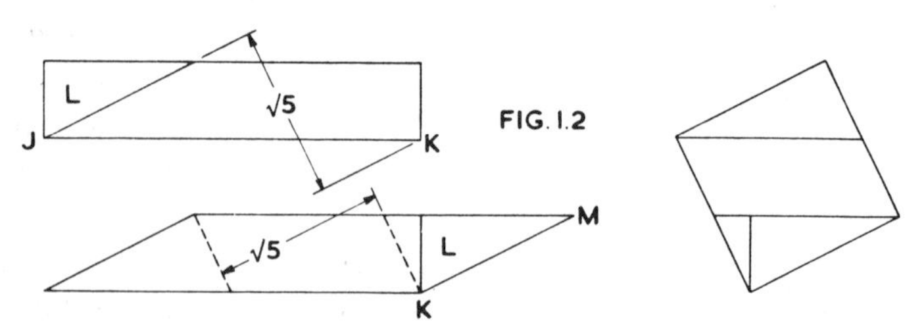

from K is $\sqrt{5}$. Transferring piece L to the right gives a parallelogram with the desired base and height, and two cuts perpendicular to *KM,* one through *K* and the other at a spacing of $\sqrt{5}$, give a dissection into a square. There is no need to go further into the various cases that can arise, for a single procedure, to be described in Chapter 2, covers all variants.

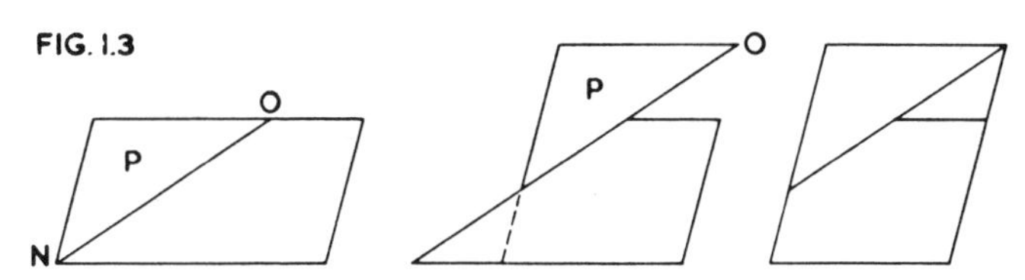

FIG. 1.3

1.2. Another kind of dissection, which I call a *P-slide,* is shown in Fig. 1.3. We make a first cut *NO,* and slide piece *P* up the incline until *O* is in line with the right-hand side of the parallelogram. We then make a second cut as shown by the broken line, put the small triangle thus obtained into the vacant space below *O,* and get a new parallelogram. Its sides are different from the corresponding sides of the original one, but its angles are the same.

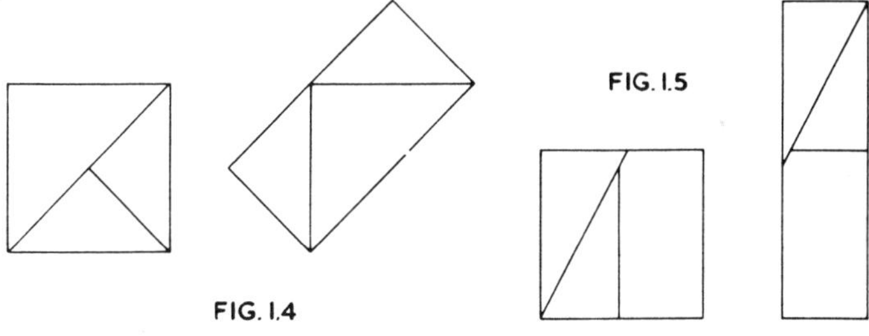
FIG. 1.4

FIG. 1.5

This fact is the great drawback of the *P*-slide, as compared with the *S*-dissection in which we can change angles. But there is one compensation, which can be made clear by dissecting a square, by both methods, into a rectangle that is as long and narrow as possible. For a three-piece dissection the cut corresponding to *AB* in Fig. 1.1 cannot exceed the diagonal of the square, so, as Fig. 1.4 shows, the *S*-dissection cannot give a rectangle in which the ratio of length to width exceeds 2:1. But Fig. 1.5 shows that with the *P*-slide the ratio can go up to 4:1. (If it exceeds 4:1, the first step would be as in Fig. 1.15 below.)

Because of this greater ratio, the *P*-slide may be superior if

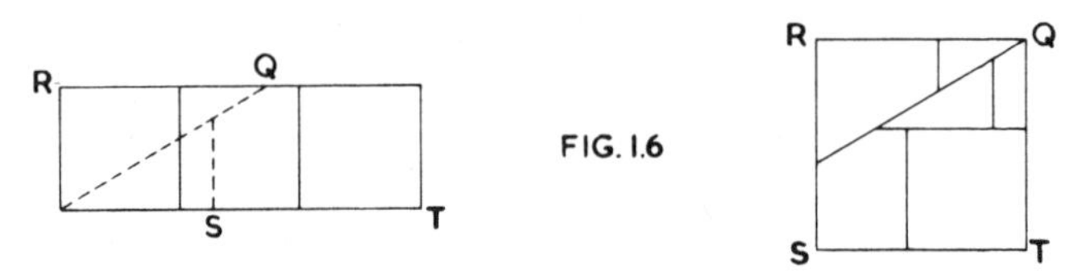

FIG. 1.6

the two polygons to be dissected have the same angles. One example is the assembly of three small squares into one large one. Putting the three squares side by side as in Fig. 1.6, we get a rectangle whose dimensions are 3 units by 1. *QR* and *ST* will become sides of the large square, and are, therefore, made equal to $\sqrt{3}$ units. This determines the positions of the cuts, which give a six-piece dissection. The *S*-dissection is less successful.

Figure 1.6 is the first of the many dissections in this book that are due to the celebrated English puzzlist Henry E. Dudeney (1857-1931). This one is a simple application of the already well-known *P*-slide; other dissections by Dudeney gave more scope for his originality.

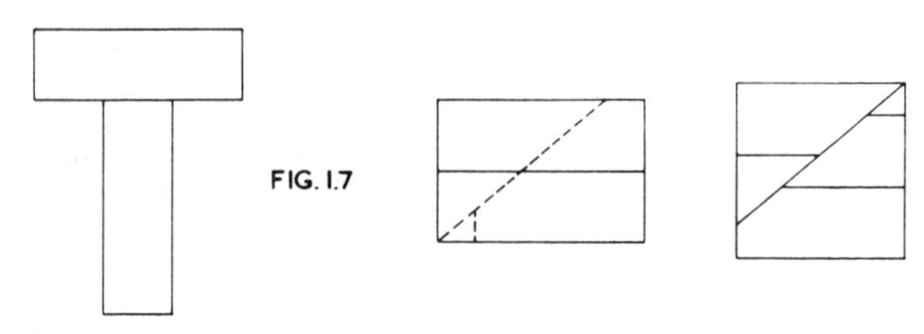
FIG. 1.7

Usually an *S*-dissection is just as good as a *P*-slide, or better. One example is Dudeney's dissection into a square of the T shown in Fig. 1.7. This T (a tau cross) has the dimensions of one obtained by cutting the edges of a hollow cube where required and laying the faces out flat.* The rearrangement into a 3 × 2 rectangle can be dissected in five pieces by a *P*-slide, as shown, but an *S*-dissection requires no more, as we shall see when we come to Fig. 2.7.

A more complex example is the assembly of three equal Greek crosses into a single square. The crosses, cut into six pieces, can be rearranged as in Fig. 1.8 into a rectangle. If the

* Can you prove that, in whatever way you do this without superfluous cuts, you cut the same number of edges?

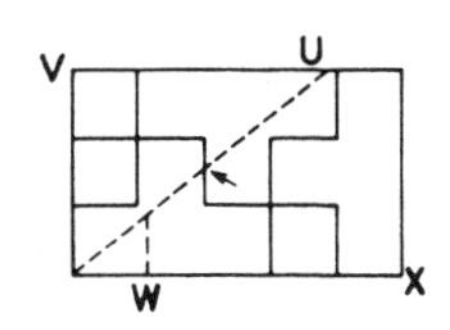

FIG. 1.8

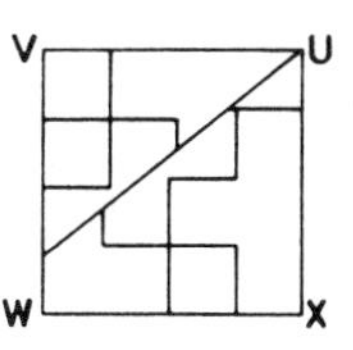

area of the cross is taken as 5 square units, the dimensions of the rectangle are 5×3, and $UV = WX = \sqrt{15}$; this locates the cuts. The dissection requires nine pieces, like the one obtained by an *S*-dissection and shown in Fig. 2.1.

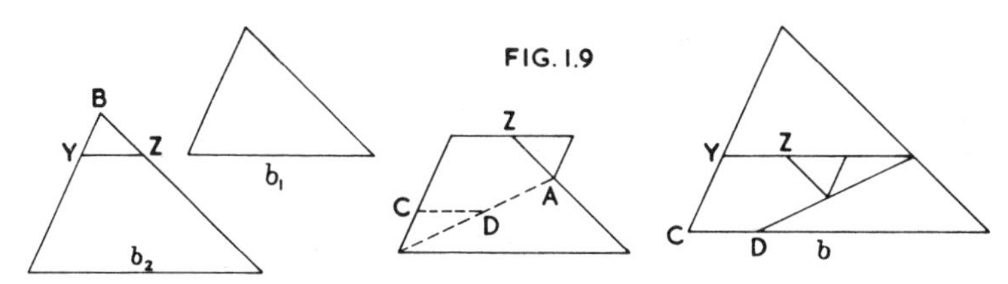

FIG. 1.9

A more subtle example of the *P*-slide is the assembly of any two similar triangles into one that is similar to them. Harry C. Bradley did this by dissecting the larger of the two into a trapezoid whose shorter parallel side is equal to the base of the smaller triangle, and putting the latter on top. If b_1, b_2, b denote the bases of the triangles with $b_1 \leqslant b_2 < b$, then it is easily found from Fig. 1.9 that $YZ = \frac{1}{2}(b_1 + b_2 - b)$, the slant cut is determined because $ZA = BZ$, and $CD = b - b_2$. Here the first step (anything but obvious) is to prepare for a *P*-slide by repositioning $\triangle BYZ$.

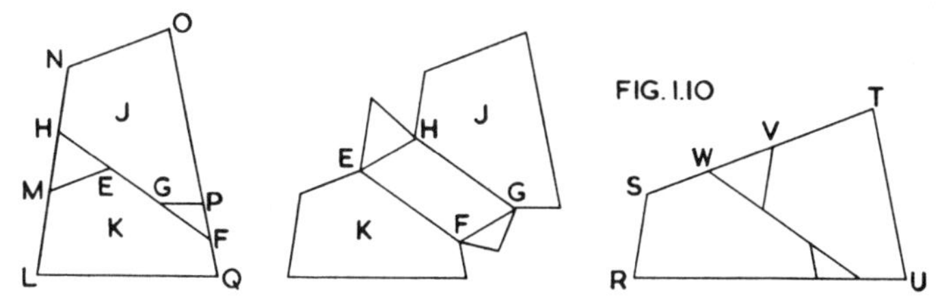

FIG. 1.10

1.3. Somewhat similar to the *P*-slide is the dissection that I call a *Q-slide*. It is of quadrilaterals, and, like the *P*-slide, it leaves angles unchanged. If the pieces of the first quadrilateral in Fig. 1.10 are hinged at E, F, G, H, then piece J, linked by the two triangles to piece K, can swing clockwise about K. Its initial and final positions give the two quadrilaterals, and examination of Fig. 1.10 shows that the cuts are determined from their dimensions as follows:

$$LM = RS,\ \ MH = HN,\ \ OP = TU,\ \ PF = FQ,\\ ME \parallel NO,\quad PG \parallel QL. \tag{1}$$

E and G must not be outside the quadrilateral, otherwise the dissection fails, but the position of E relative to G is immaterial. We see this from Fig. 1.11, the dissection of a triangle (a quadrilateral in which one side has zero length) into a trapezoid. Here the initial order is $HGEF$, whereas in Fig. 1.10 it is $HEGF$.

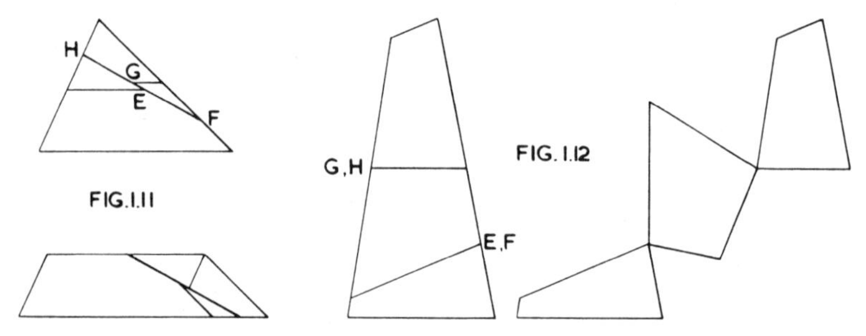

FIG. 1.11

FIG. 1.12

Figure 1.11 provides an application of the Q-slide, namely a dissection of triangles alternative to Fig. 1.9. Here again the smaller of the two is undissected. This dissection is more general than that of Fig. 1.9, in that it can be applied to trapezoids as well as triangles.

As the proportions of the quadrilaterals vary, E in Fig. 1.10 can become lower than G, as in Fig. 1.11, and in the limiting case E coincides with F. At the same time G coincides with H. In this limiting case the transverse cut HF is unnecessary, as is seen from Fig. 1.12. This is a feature in common with the P-slide, as a glance at Fig. 1.5 shows. Another common feature is that angles are unchanged. Nevertheless, the differences between the two slides are sufficient for them to be considered distinct.

1.4. There is a further kind of dissection, which is quite popular, given the descriptive name *step dissection*.

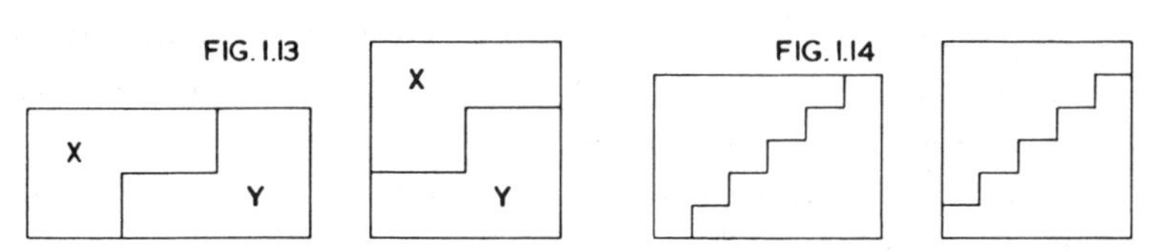

The simplest step dissection is that by which a 9×4 rectangle is changed to a square, not in as many as three pieces, but only in two. In Fig. 1.13 piece *X*, moved up a step over piece *Y*, gives the square. A dissection with more steps is shown in Fig. 1.14. It can be seen that in these two-piece dissections the rises of the steps must all be equal, and so must the treads.

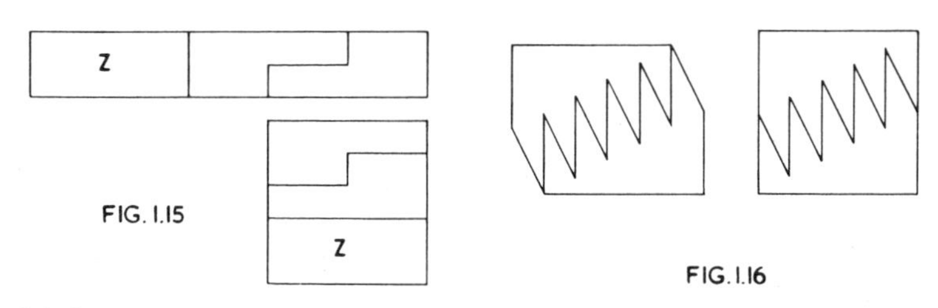

Slightly more complex is the dissection of a 6.25×1 rectangle into a square. Since 6.25 is greater than 4, an *S*-dissection or a *P*-slide in three pieces is impossible. But it can be done with a step dissection. As shown in Fig. 1.15, we first cut off a rectangle *Z* whose length is equal to the side of the equivalent square, namely 2.5, and the remainder, of dimensions 3.75×1, is dissected into a rectangle of dimensions 2.5×1.5. This is undoubtedly the second longest and narrowest rectangle that has a three-piece dissection into a square.*

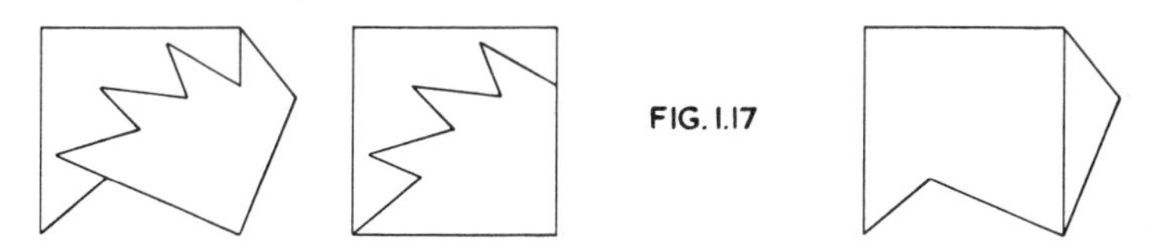

As is seen from Fig. 1.16, the steps need not be rectangular. Nor is it necessary that the relative displacement of the two pieces be a translation, i.e., no rotation; there can be relative rotation, as in Fig. 1.17. (This dissection is more elaborate than

* What is the longest and narrowest?

the alternative on the right. But Dudeney, who originated it, added a condition to his puzzle to exclude the simpler alternative.) In several dissections of this kind the rotation is a quarter turn. One example by Dudeney is shown in Fig. 1.18.

FIG. 1.18

It will be realized that a step dissection imposes severe restrictions on the dimensions of the polygons to which it is applied. For instance, two-piece dissections as in Figs. 1.13 and 1.14 require the dimensions of the rectangle to be $n^2 \times (n + 1)^2$ where n is an integer, the side of the equivalent square being $n(n + 1)$. And the three-piece dissection of Fig. 1.15 is not applicable to a rectangle of width 1 unit, if its length is 6.24 or 6.26. Contrasting with this, the dimensions of polygons can be varied continuously without precluding an *S*-dissection or a slide. In general a step dissection requires that there be a rational ratio between sides or angles in the one polygon and sides or angles in the other. If we call such dissections *rational*, we have a niche in our classification system (*R-dissections*) for further dissections that would otherwise be left out in the cold. Two of them, both assemblies of two squares into one, are shown in Figs. 1.19 and 1.20. Usually dissections like these are made necessary by some restriction; e.g., the polygons bear a chessboard or patchwork pattern that must be preserved.

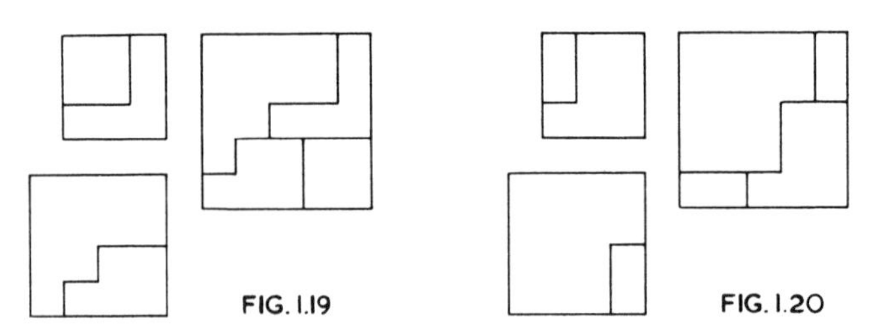
FIG. 1.19 FIG. 1.20

1.5. All the dissections described are reversible, in the sense that the procedure is in substance the same whether we dissect polygon A into polygon B or vice versa. For instance, in dissecting the second quadrilateral of Fig. 1.10 into the first, the cuts are determined by $TV = ON$, $VW = WS$, and so on, the equalities and parallelisms severally corresponding to those in (1). Partly because of this reversibility, I shall also refer to a dissection as that of polygons *A and B*.

S-dissections and the slides appear to be the only dissections that do not depend on any lines in the respective polygons being equal or in some rational ratio. But there may be others awaiting discovery. After all, so far as I know the *Q*-slide was discovered as recently as 1950 or thereabouts, and no dissection using it was published before 1957.

The *S*-dissection has as yet been given scant treatment. Actually it is by far the most useful. In Chapter 2 we shall make up for the neglect, and show how versatile it is.

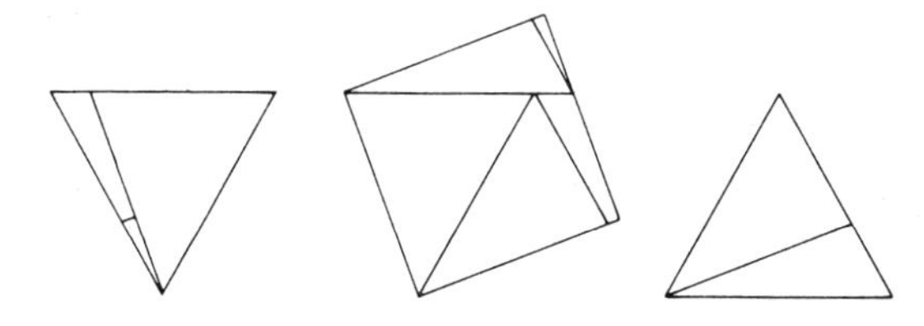

2

P-STRIPS AND PP DISSECTIONS

In Fig. 1.1, piece C was detached from the left-hand side of the parallelogram and placed on the right, with AH lying along DE. Now there is no need whatever for AH and DE to be straight lines; the join between A and H can take any course we like that lies wholly between the parallel lines AD and HE, provided only that it is identical with the join between D and E. An S-dissection is then possible.

FIG. 2.1

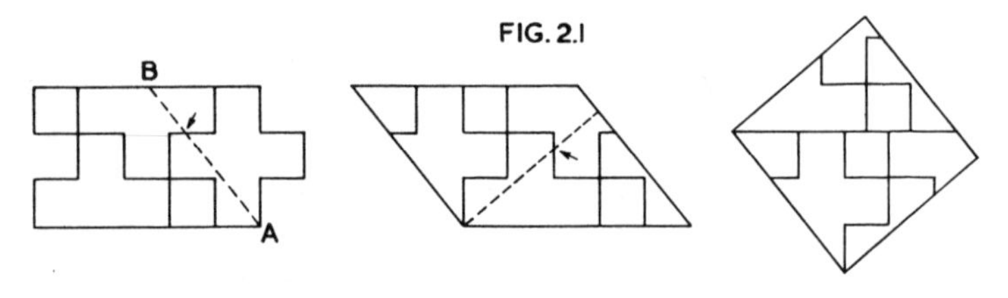

This can be illustrated by a dissection of three equal Greek crosses into a square, alternative to Fig. 1.8. Cutting the crosses into only five pieces instead of six, we can assemble them as shown on the left of Fig. 2:1. The right and left sides of the assembly fit together like mortise and tenon, so an S-dissection is possible. In applying it we make the first cut AB beginning at the lower right vertex, not at the lower left one, simply because trial shows that we thereby get a minimal dissection. (This matter of trial will be taken up presently.)

Apart from the trivial difference that A is on the right, the dissection in Fig. 2.1 is substantially the same as in Fig. 1.1, and requires nine pieces like the one in Fig. 1.8. The equality is easily accounted for. The initial set-up in Fig. 2.1 has five pieces, and so is one better than that of Fig. 1.8. But in completing the dissection in Fig. 2.1, the added cuts cross existing

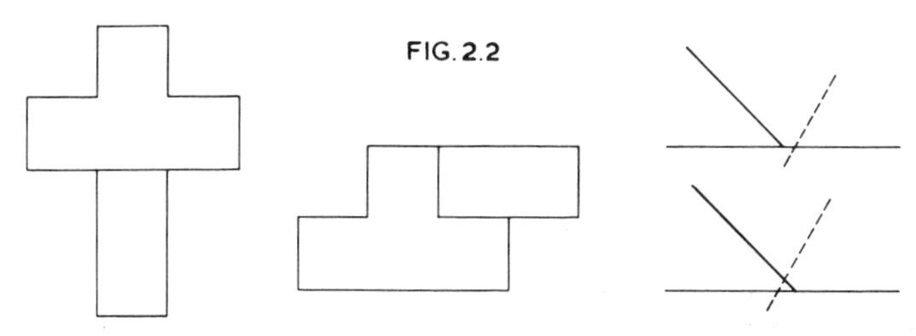
FIG. 2.2

ones in two places, indicated by arrows; in Fig. 1.8 an added cut crosses an existing one in only one place, also indicated by an arrow. Now each point where cuts cross means an extra piece. Therefore, the added cuts in Fig. 2.1 (with two crossings as against one) cancel the initial advantage.

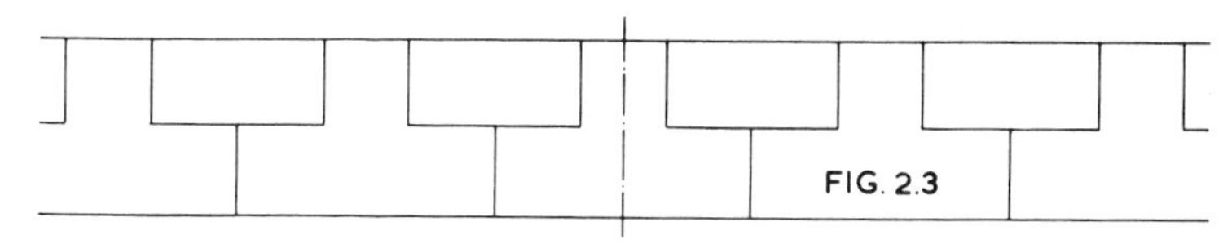
FIG. 2.3

This trouble of cuts that cross crops up in all dissections. Just as in Fig. 2.1, so in other dissections cuts *will* cross, and our problem is to find a dissection in which the points where cuts cross are as few as possible. Now the possibilities are endless, even though the cut like *AB* in Fig. 2.1 must be at a fixed angle. For point *A* of the cut need not be at a vertex, but can be anywhere on the base; *AB* can slope the other way; and the second cut, like the first, need not start at a vertex, and can slope the other way. To visualize all the possibilities from the figures is out of the question if they are at all complicated; to draw them all would take an age. Fortunately the trial becomes quite speedy and quite a pleasure if we make use of *strips* in the manner now to be described.

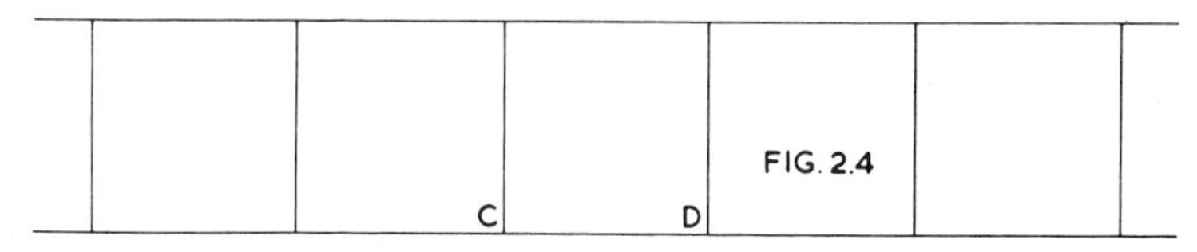

FIG. 2.4

Let us turn for a change from Greek crosses to the Latin cross illustrated in Fig. 2.2. Cutting it into two pieces, we can assemble them into a figure to which an *S*-dissection can be applied, because the joins on left and right fit together. Well then, let us take three or four figures and fit them together as in Fig. 2.3. Do the same, on a separate sheet of tracing paper, with the polygon you want to dissect the Latin cross into; if it is a square, you get Fig. 2.4. Next, superpose the strip of squares on the strip derived from crosses at an angle such that *C* and *D* are respectively on the lower and upper parallel lines in Fig. 2.3. Figure 2.5 shows the sort of thing you might get, with broken

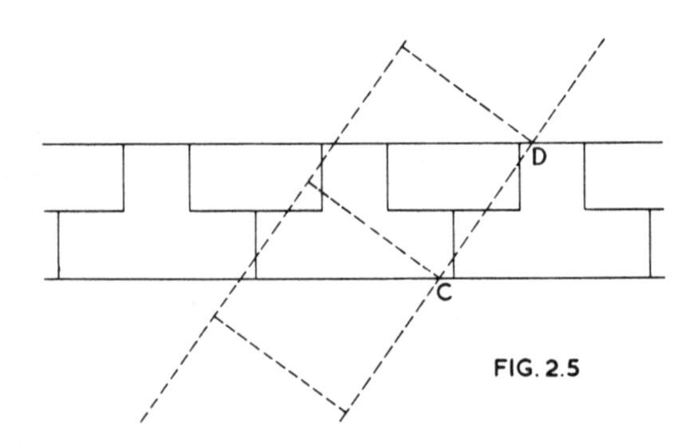

FIG. 2.5

lines crossing full ones at several points. Slide the strip of squares relatively to the other strip, keeping the angle between the two constant, until the number of crossing points is reduced to the minimum. It will be found that the minimum is 1, as in the three superpositions shown in Fig. 2.6. Now look at the parallelo-

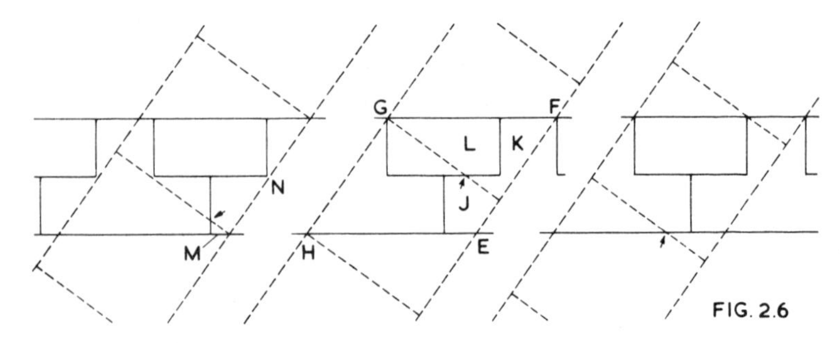

FIG. 2.6

gram *EFGH* in the second one. If we move pieces *J* plus *K* to the left so that line *EF* thereof lies along *HG*, we get a figure which is immediately changeable to a Latin cross, as on the left of Fig. 2.7. If we move pieces *K* plus *L* downwards so that line *GF* thereof lies along *HE,* we get a square as on the right of Fig. 2.7. Thus, the lines in the area *EFGH* common to the strips give a dissection of Latin cross and square. Making the strip of squares slope the other way does not in this case give different dissections but only mirror images, since the strip derived from crosses has centerlines of symmetry like the one shown in Fig. 2.3, and so has the strip of squares.

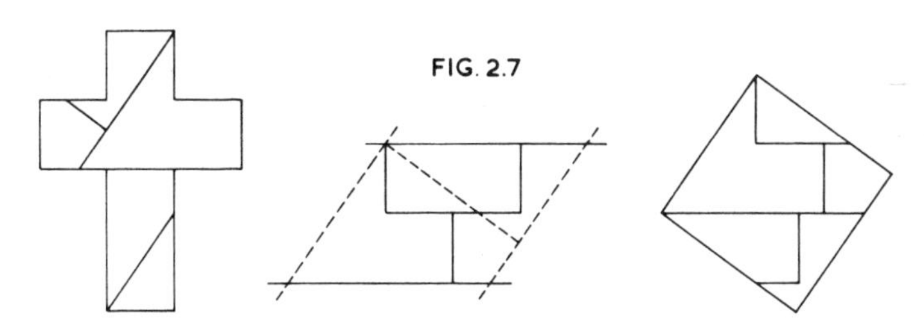
FIG. 2.7

The strip element in Fig. 2.2 can be formed into the tau cross in Fig. 1.7. Thus Fig. 2.7 is an alternative to Fig. 1.7.

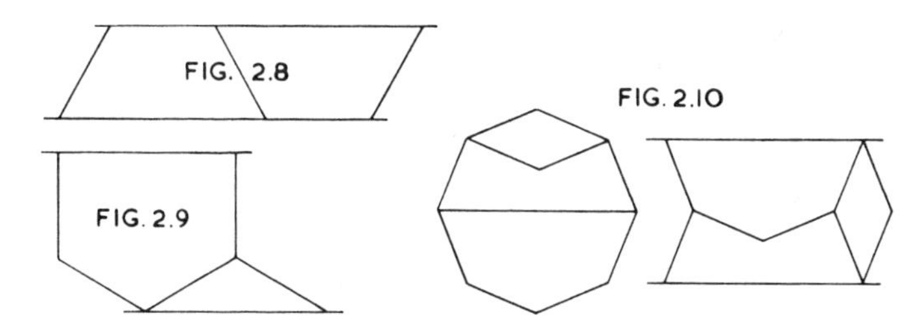
FIG. 2.8

FIG. 2.9

FIG. 2.10

This method of superposition will give a dissection of any two polygons whatever. In brief, cut up each polygon if necessary so as to make it a strip element, superpose the strips at the correct angle, and the lines in the common area give a dissection. A random superposition is not likely to give a minimal dissection; thus, Fig. 2.5 requires eight pieces. But the best one obtainable from the strips used is easily found by trial. Whatever their relative position in the best dissection or any other, the edges of the one are similarly located relatively to the elements of the other in which they lie. In previous articles on the subject, points such as those marked by dots in Fig. 2.4 have been called *congruent*. One can say then that the strips are superposed so that the edges of each pass through congruent points in the other. And the *S* in "*S*-dissection" indicates that it is obtained from Strips.

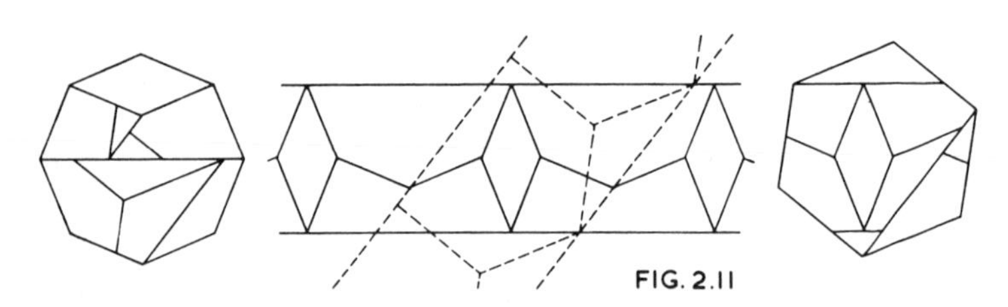

FIG. 2.11

The elements of the strips considered so far have a parallelogram as prototype. For this reason they are called *P-strips,* to distinguish them from another kind of strip that can be used. When one *P*-strip is superposed on another, the result is called a *PP dissection.*

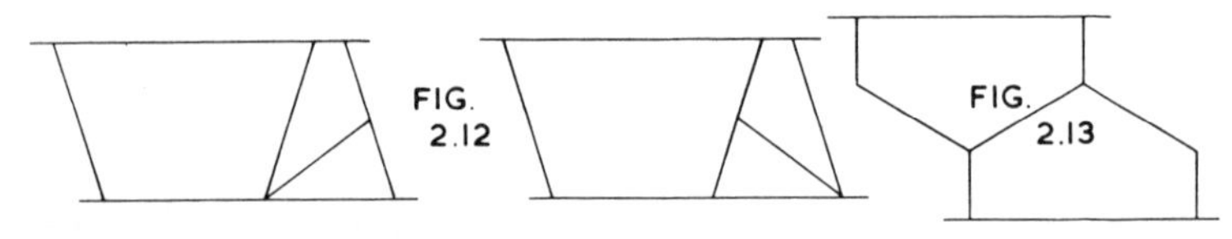

FIG. 2.12 FIG. 2.13

Most of the simpler polygons one might want to dissect are easily made into strip elements, and often in more than one way. When the latter is the case, the various simple ways of getting strip elements should be tried. For it sometimes happens that a minimal dissection is not given by the simplest strip ele-

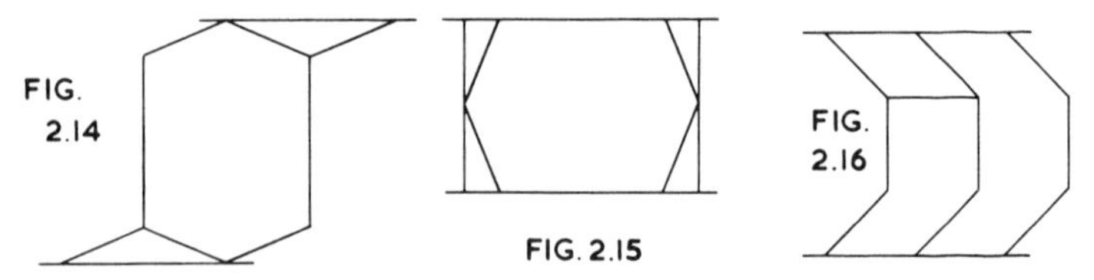

FIG. 2.14 FIG. 2.15 FIG. 2.16

ment, or it is given by one element but not by another that is equally simple. Thus, there is not much to choose between the strip elements in Figs. 2.8 and 2.9, both derived from a hexagon. But only the second will give a nine-piece dissection of octagon and hexagon, when superposed on a strip with element as in

Fig. 2.10. The superposition and the dissected polygons are shown in Fig. 2.11.

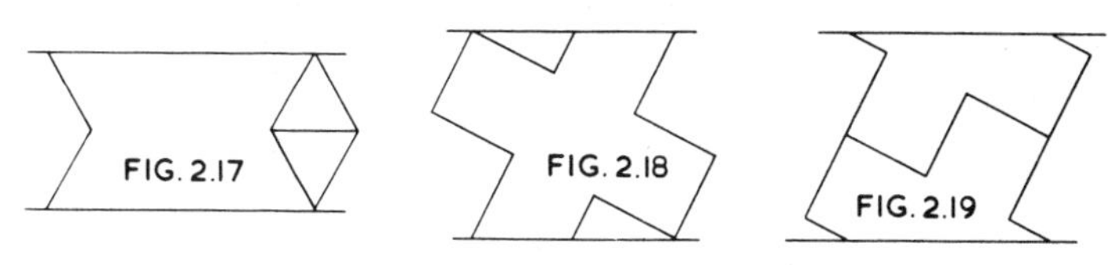
FIG. 2.17 FIG. 2.18 FIG. 2.19

Some simple ways of dissecting various polygons into *P*-strip elements are shown in Figs. 2.12-2.21. The pentagon strip elements in Fig. 2.12 are obtained by first getting a trapezoid as in Fig. 2.22 and making a cut parallel to the left-hand side, through the midpoint marked by a dot. Fig. 2.22 also explains Figs. 2.16 and 2.17. The rest of Figs. 2.12-2.21 hardly need explanation.

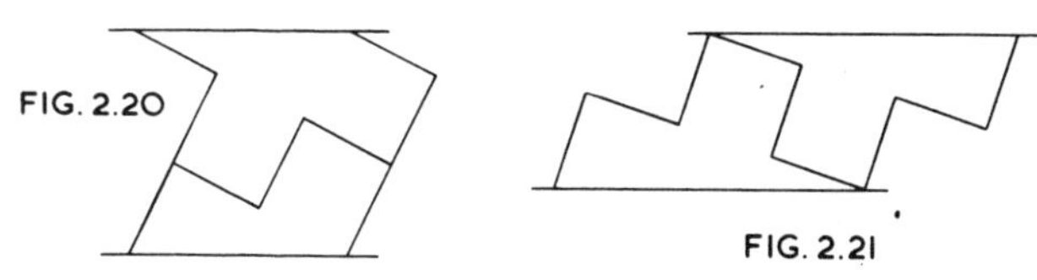
FIG. 2.20 FIG. 2.21

I have found only one application of Fig. 2.16. This is the dissection of octagon and golden rectangle. (The latter is the rectangle in which length/width $= (1 + \sqrt{5})/2$. See Martin Gardner, *Second Scientific American Book of Mathematical Puzzles and Diversions,* Chapter 8.) The unique application is shown in Fig. 2.23.

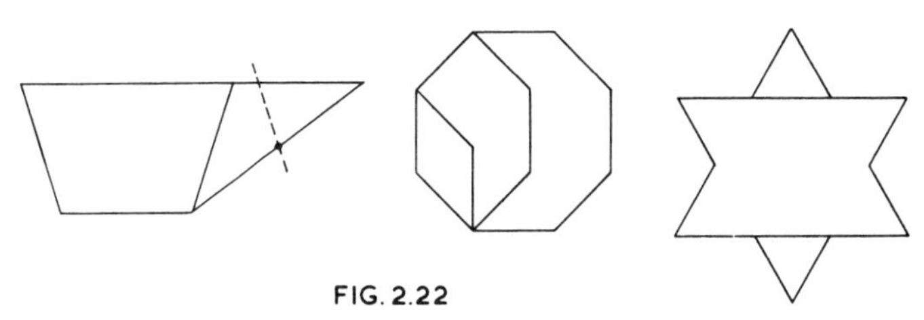
FIG. 2.22

You are now in a position to appreciate the value of proceeding as follows:

(1) Decide on a fixed area for all polygons. (I found 20 sq. cm. convenient.)

(2) Make accurate drawings on tracing paper, preferably in

India ink. (It is annoying to find that you are uncertain whether two lines cross or not, solely because your drawings are not accurate enough. The sketches on the right of Fig. 2.2 illustrate the dilemma.)

(3) Keep all drawings.

You will then get better results, and the saving in time and trouble will be *enormous*.

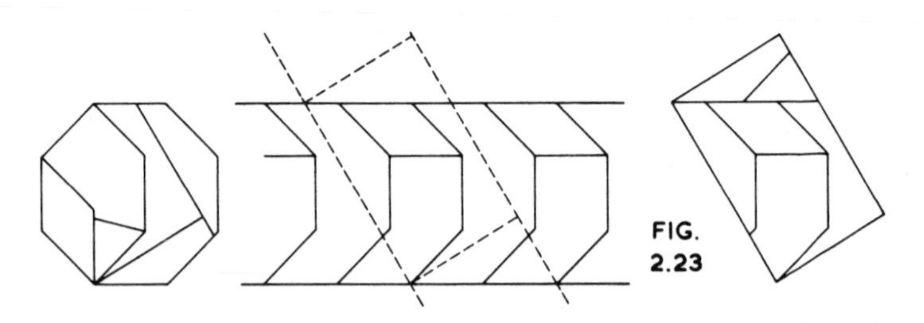
FIG. 2.23

An incidental advantage of using tracing paper is that if two strips are mirror images, only one need be drawn—just turn it over and you get the other. This applies to the pentagon strips with elements as in Fig. 2.12. Figure 2.16 also is one of a pair, and so is each of the Greek-cross elements in Figs. 2.18-2.21.

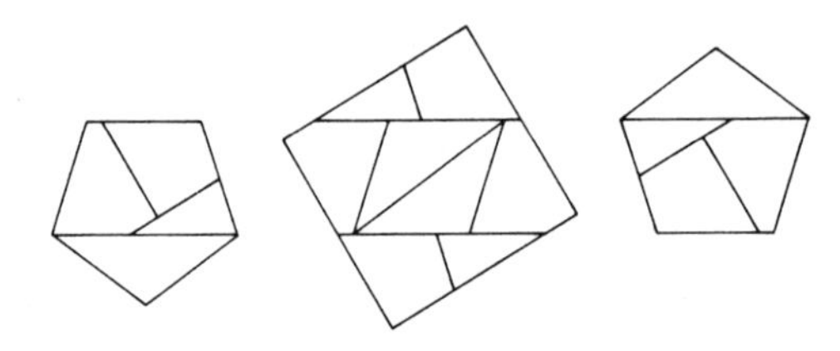

3

MORE *PP* DISSECTIONS

A few minimal *PP* dissections have been described in Chapters 1 and 2. In this chapter, I present several more of this kind, obtained from strips already illustrated.

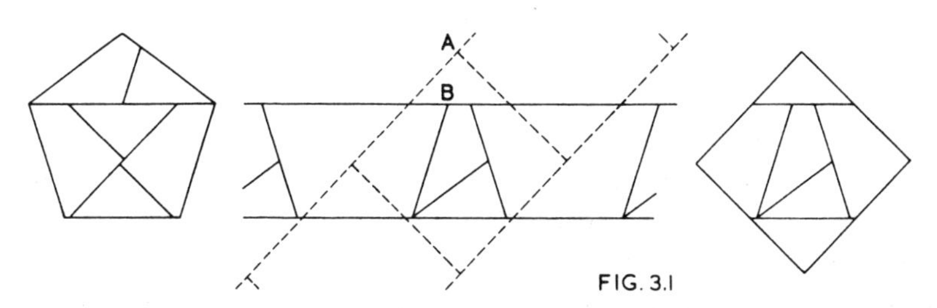

FIG. 3.1

The dissection of pentagon and square in Fig. 3.1 is similar to one discovered by Dudeney. His dissection has *A* in the strip of squares coincident with *B* in the pentagon strip. But in the dissection shown the pieces are more of a size.

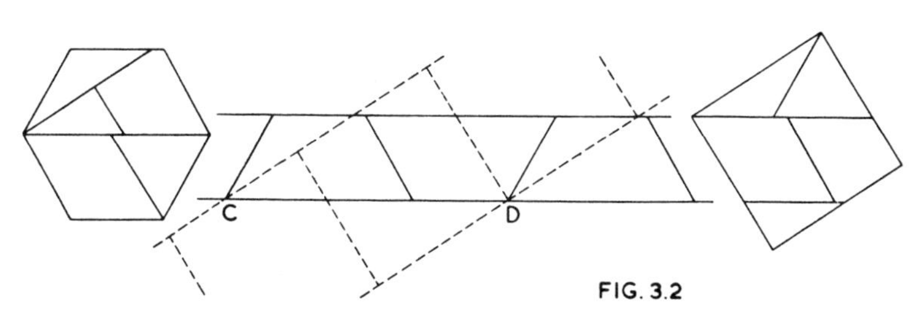

FIG. 3.2

It will be recalled that in comparing the *S*-dissection and *P*-slide by means of a dissection of square and rectangle, the length of the first cut in Fig. 1.4 was limited to that of the diagonal of the square. But this was for a three-piece dissection. Actually the cut can be as long as we like, and in Fig. 3.2, also Dudeney's, *CD* does in fact exceed the diagonal. The longer the cut, the greater the number of pieces is likely to be. But they can always be found by superposing strips.

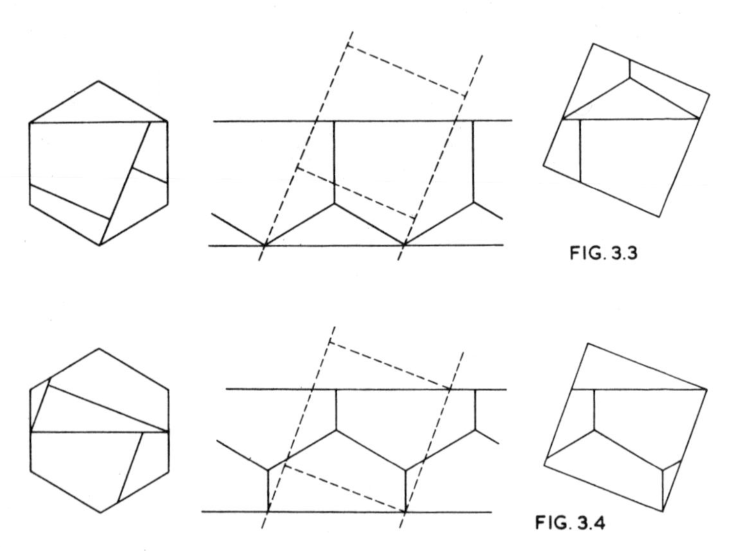

FIG. 3.3

FIG. 3.4

On the other hand *CD* has a minimum length below which superposition is not possible, for it must reach from edge to edge of the other strip. It can fall below the minimum if both strips are rather wide. The only pair of strips already illustrated for which this occurs consists of those in Figs. 2.14 and 2.16. Trial will show that one cannot dissect an octagon into an octagon (!) using these two strips, for they cannot be superposed in the prescribed manner.

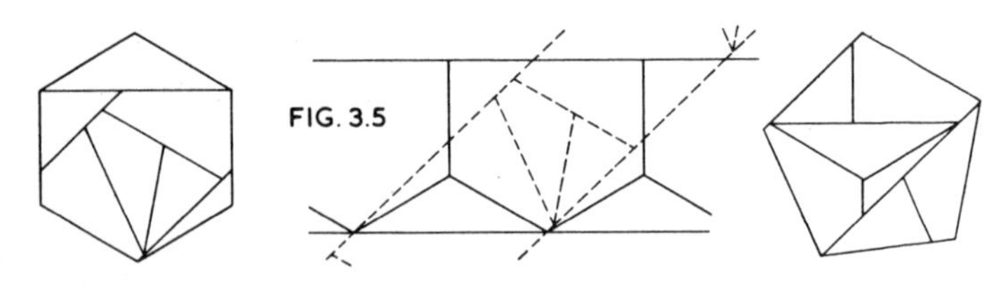

FIG. 3.5

In Figs. 3.2-3.4, we have three different five-piece dissections of hexagon and square. When there are several different dissections of a given pair of polygons, as here, the question arises, which is "best"? Inspection of Fig. 2.6 suggests two criteria which, I think, will meet with approval. The first is that very small pieccs such as *M* are a blemish. The second is that pieces with a weakness such as the isthmus *N* are likewise a blemish. In addition a very narrow piece, e.g., a triangle whose height is 100 times its base, would be undesirable. By these criteria the second dissection in Fig. 2.6 is the best, and, by a slight margin, Fig. 3.2 is better than Figs. 3.3 and 3.4.

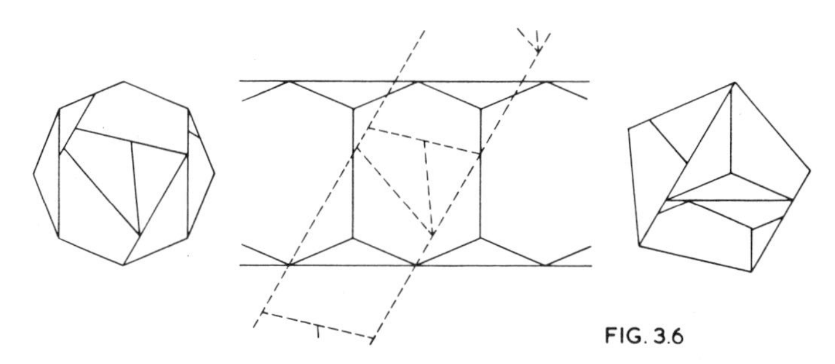

FIG. 3.6

Figures 3.5 and 3.6 show how strips containing one large piece are more likely to give economical dissections. When the strips are superposed, it is more likely that small pieces in each can be placed wholly within the large piece in the other, reducing the number of cuts that cross. The hexagram dissection in Fig. 3.9 below illustrates the same point.

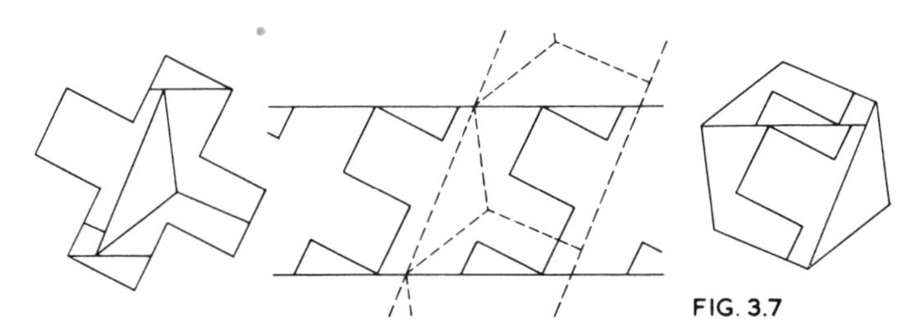

FIG. 3.7

Another seven-piece dissection of Greek cross and hexagon, alternative to Fig. 3.7, is obtained with the hexagon strip sloping the other way. It is most unbeautiful, but a use will be found for it in Chapter 22.

The six-piece dissection in Fig. 3.8 is obtained by quite a close shave; if *EF* were a little greater than *EG* instead of a little less, an extra piece would be required. A reasonably accurate

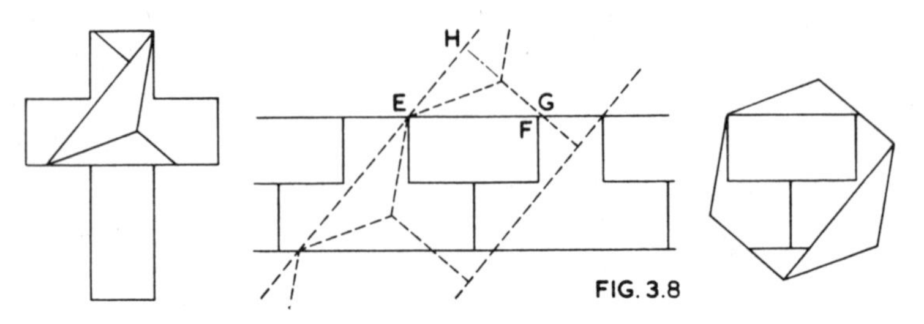

FIG. 3.8

drawing will show the correct position of F relative to G, but for reassurance we must ascertain it by calculation. For a polygon area of 20 square units, it will be found from the data in Appendix C that the angle between strips is

$$\sin^{-1} 3.6514/4.8056 = 49° \ 27',$$

whence

$$EG = EH \sec \angle HEG = 2.4028 \sec 49° \ 27' = 3.6959.$$

On the other hand $EF = 3.6514$, which, to our satisfaction, is less than EG.

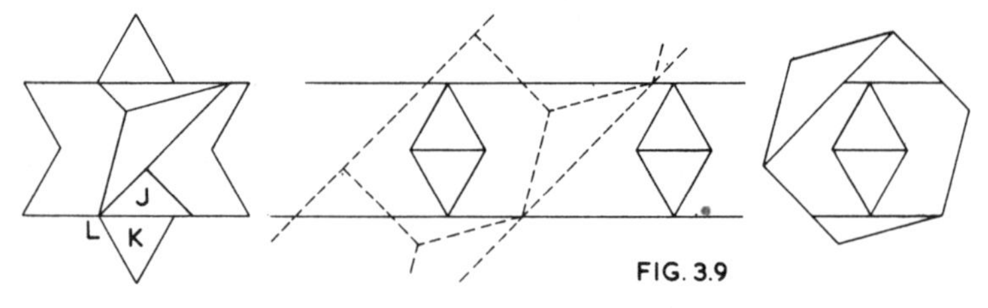

FIG. 3.9

Often the relative position of the strips in a *PP* dissection can be varied appreciably, changing the proportions of the pieces but not the nature of the dissection. In Fig. 3.9 the hexagon strip can be displaced horizontally over a span that is unusually large, without chipping the diamonds. It has been positioned in the figure so that pieces J and K have a common vertex L.

The dissection in Fig. 3.10 contains some ungainly pieces. But I have found no other dissection of Greek cross and hexagram that requires only eight pieces. This property outweighs all the elegancies!

It is desirable to distinguish the two strips in a drawing of their superposition. Therefore broken lines are used for the

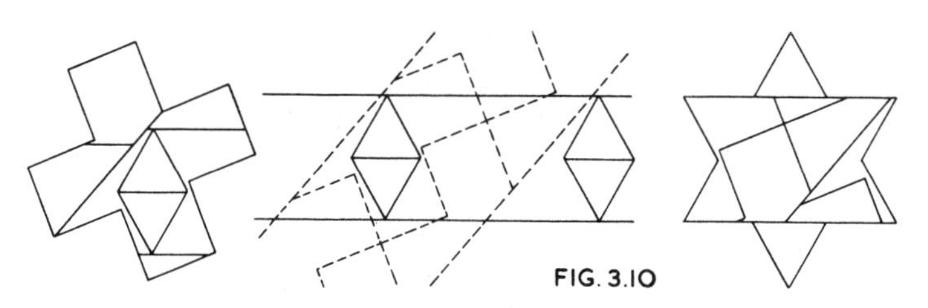

FIG. 3.10

sloping strip in every drawing of a superposition in this book. This harmonizes with the practice, adopted from Fig. 1.1 onwards, of indicating added cuts by broken lines. And the dissected polygons, appearing on the left and right of each superposition, are oriented so as to make it as easy as possible to identify the pieces therein with those in the superposition.

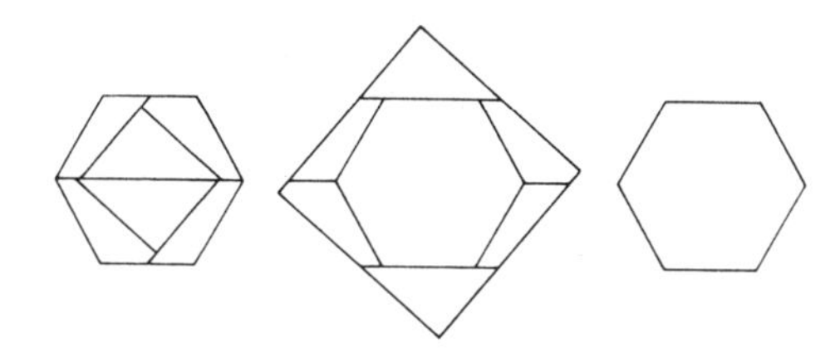

4

T-STRIPS AND *PT* DISSECTIONS

In making the pentagon into a strip element, the first step was to change it into a trapezoid, as shown in Fig. 2.22. But this first step already gives us a strip element, as is seen from Fig. 4.1. The strip differs from those used hitherto in that alternate elements are inverted. Being composed of trapezoids, it is called a *T-strip*.

Such a strip also can be used to get a dissection, but we cannot translate the strip superposed on it as freely as on a *P*-strip. A strip of squares, when superposed as in Fig. 4.1, does not give

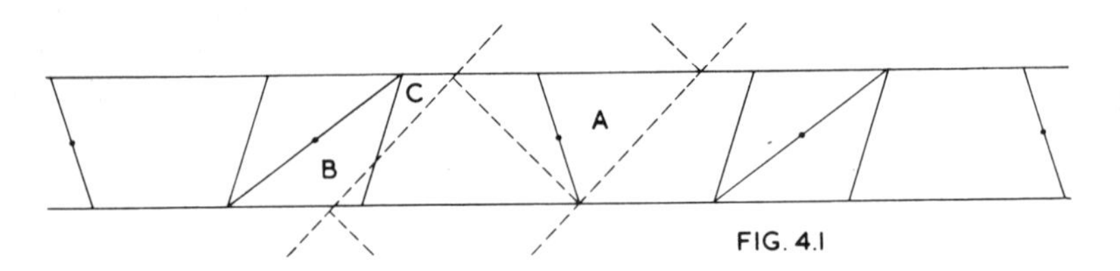

FIG. 4.1

a dissection, for the common area contains a piece *A* that does not fit over *B* plus *C* as required for the pentagon. But suppose the edges of the superposed strip pass through adjacent mid-points (shown by dots), as in Fig. 4.2. The common area, which

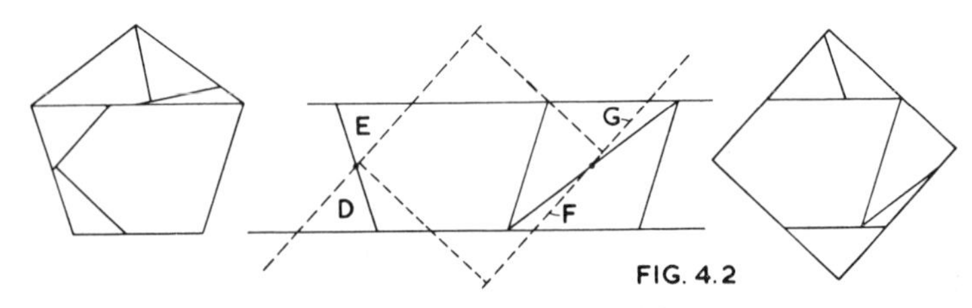

FIG. 4.2

is clearly a dissection of a square, can be made into a pentagon by swinging piece *D* up to *E*, and piece *F* up to *G*. Hence, a *T*-strip gives a dissection just as easily as a *P*-strip, provided that the edges of the other strip pass through midpoints of the internal edges of the *T*-strip element. These midpoints are also centers of symmetry of the strip, whence pairs of pieces like *D* and *E* are identical, and we have a dissection. Because of the role played by the midpoints of *T*-strips, they will always be marked by dots.

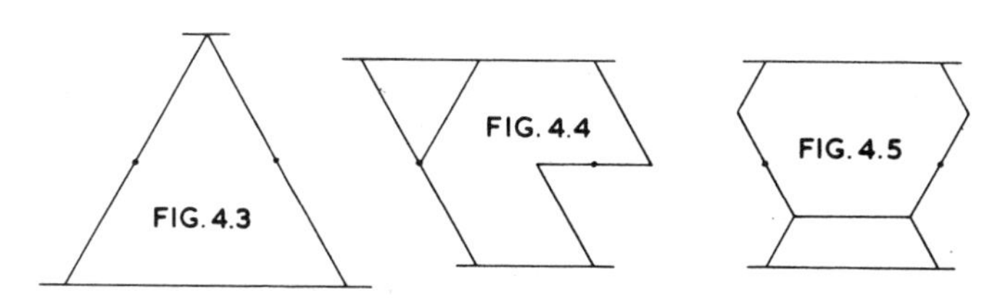
FIG. 4.3 FIG. 4.4 FIG. 4.5

Just as a parallelogram is only the prototype of a *P*-strip element, so a trapezoid is only the prototype of a *T*-strip element. More generally, the nonparallel sides of the trapezoid can be replaced by any joins, lying wholly between the parallel sides, such that each has a midpoint about which it is symmetric. This applies to the rearrangements of hexagon and Greek and Latin crosses in Figs. 4.4-4.7, which are, therefore, *T*-strip elements.

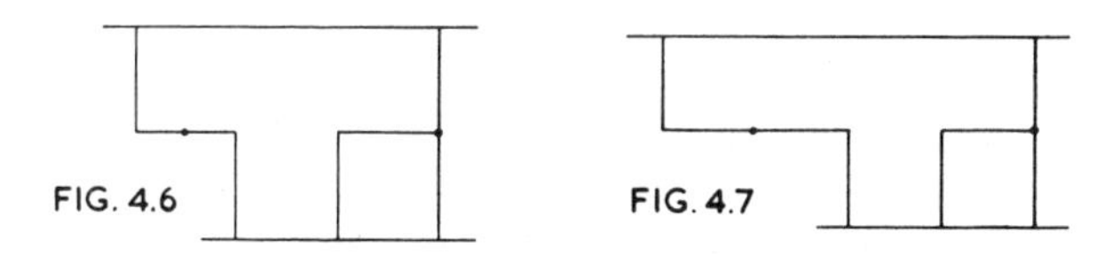
FIG. 4.6 FIG. 4.7

Some *PT* dissections obtained from strips already illustrated are shown in Figs. 4.8-4.11.

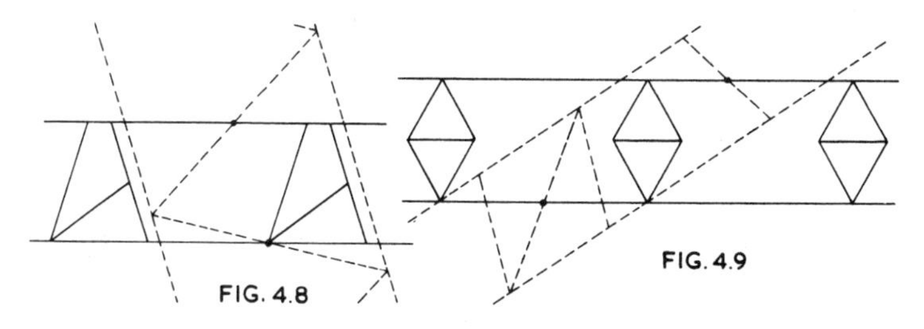
FIG. 4.8 FIG. 4.9

Another kind of strip, too uncommon to merit a separate

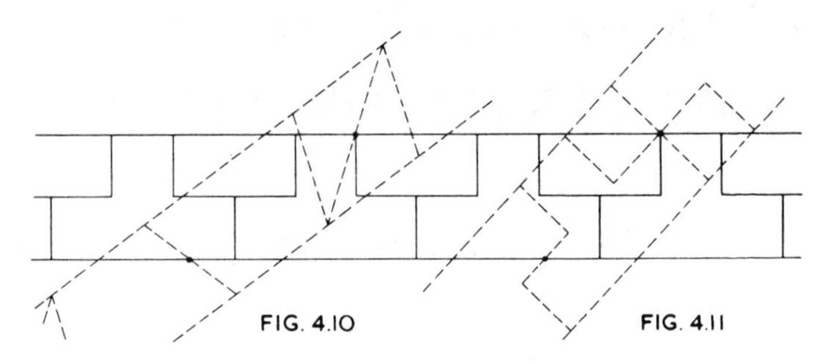
FIG. 4.10 FIG. 4.11

chapter, has alternate elements turned over. The square plus triangle on the left of Fig. 4.12 is made into the element of such a strip by the single cut shown, and a three-piece dissection into a square is obtained by superposing a strip of squares with the strips coincident or perpendicular. Because alternate strips are turned over, one of the pieces in the dissection must be turned over too. In general, for coincident superposition, the strips must have the same width, and both must have alternate elements turned over. The alternative is superposition with the strips perpendicular. For this the mean length of each strip must be equal to the width of the other, but it is not necessary that both strips have alternate elements turned over.

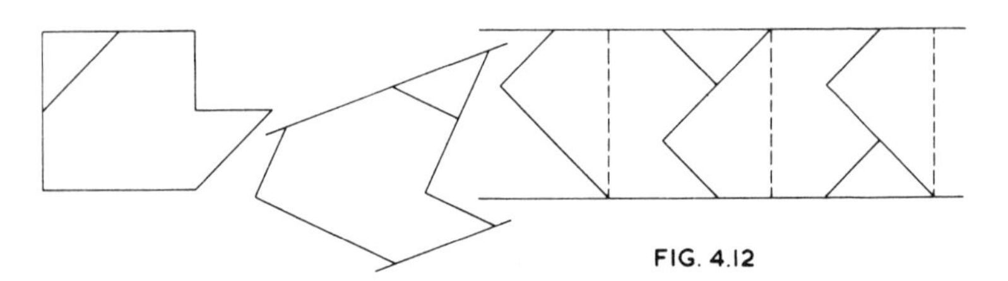
FIG. 4.12

Figure 4.12 was published by Sam Loyd (1841-1911), Dudeney's American counterpart, who had an unsurpassed gift of devising problems that seem insoluble. His fertility of invention extended to every kind of problem, including the chess problems which are, I should say, his most durable monument.

From now on the rearrangements of the common area into the two polygons will usually not be illustrated. By now you should not need such spoonfeeding.

5

TT DISSECTIONS

If a T-strip has another strip superposed on it, the edges of the latter must pass through midpoints of the T-strip. If both are T-strips, this requirement works both ways, that is, the edges of each must pass through midpoints of the other. Figure 5.1 shows an example.

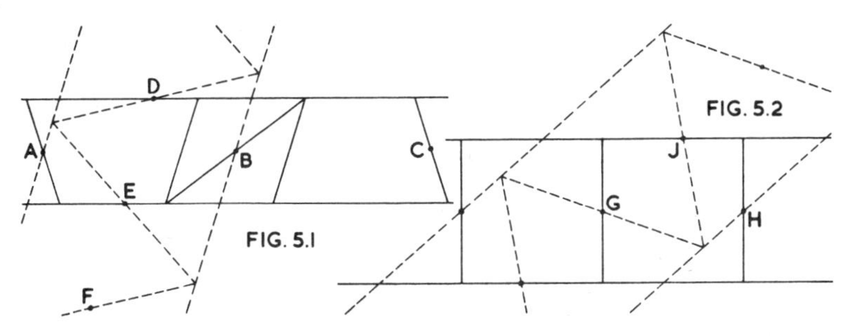

In consequence of this, we cannot slide either strip continuously over the other; only a finite number of TT dissections are possible with two given strips. They are all obtained by passing the edges of the second strip through B and C in the first as well as through A and B, passing the edges of the first through E and F in the second as well as through D and E, and ringing the same changes with the oblique strip sloping the other way and with one strip turned over. The number of different dissections thus obtained depends on the asymmetry of the strip elements. It is usually found that one and one only of the dissections is appreciably better than the others.

Because of the severe restrictions on the relative positions of the strips, you might think that TT dissections are not much use. On the contrary, several minimal dissections are of this kind, including Dudeney's famous one illustrated in Fig. 5.2.

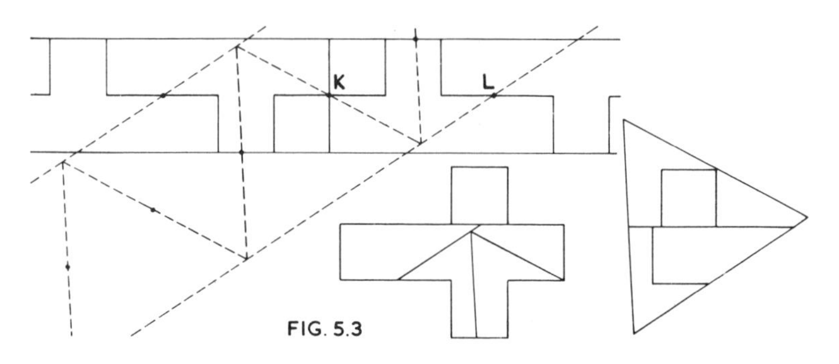

FIG. 5.3

Most TT dissections differ from that of Fig. 5.1 in tnat the common area is not equal to that of each polygon, but is twice as much. Typical dissections of this kind are shown in Figs. 5.2 and 5.3. The edges of each strip, as can be seen, do not pass through adjacent points in the other; instead (referring to Fig. 5.3) there is a midpoint K in between, where lines in both strips intersect at the center of the common area. The latter in effect contains a dissection of two crosses into two triangles. But as K, being a common midpoint, is a center of symmetry for each strip, it is also a center of symmetry for the common area. Hence, the pieces in half the common area give a dissection of one cross into one triangle. The half can be one that is bounded by any join, going from edge to edge of either strip, that is symmetric about K. The kinds of dissection in Figs. 5.1 and 5.3 are distinguished by the respective names $TT1$ and $TT2$.

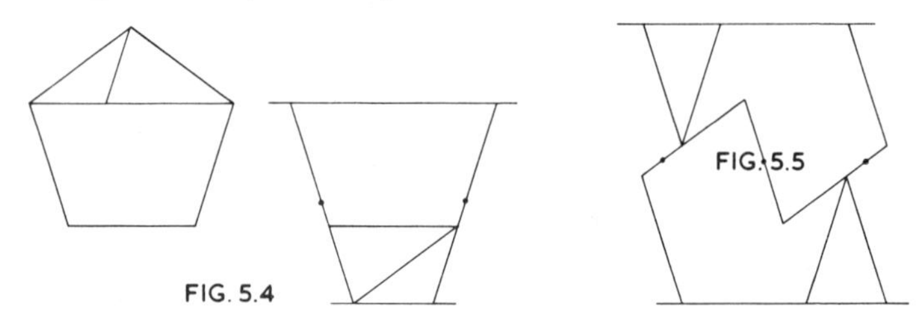
FIG. 5.4

FIG. 5.5

Figure 5.3 will make clear one of the advantages of $TT2$ superpositions. The width of one strip may be greater than the mean length of the element of another; e.g., in Fig. 5.3 the width of the triangle strip is greater than KL. In such cases superposition would normally be impossible. But if both are T-strips, $TT2$ provides a way out.

The strip of squares used in Fig. 5.2 has hitherto been called a *P*-strip, but it is clearly a *T*-strip too. In a *TT*2 dissection it must be superposed in the *TT*2 way, providing a transverse line through *G* to divide the common area into two halves.

Several other polygons besides the Latin cross and square have a *TT*2 dissection into a triangle, as you will find from Fig. 15.1 and Appendix D. Those of the hexagon, heptagon, octagon, and Greek cross are so similar that they look like mass production!

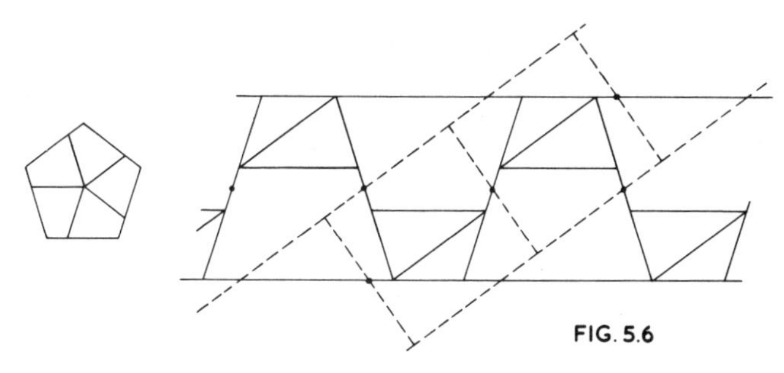

FIG. 5.6

There are at least two other ways besides Fig. 4.1 in which a *T*-strip can be obtained from a pentagon. They are shown in Figs. 5.4 and 5.5. In the latter, the strip element does not reach right across from edge to edge, and for this reason two elements are shown. The first element is used in Irving L. Freese's dissection of pentagon and square, shown in Fig. 5.6.

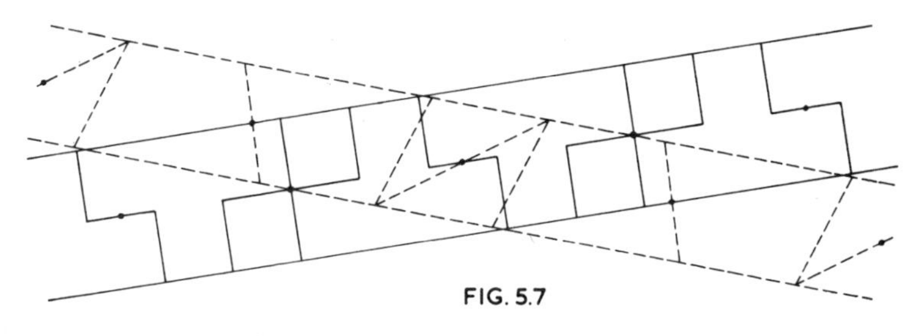

FIG. 5.7

Figure 5.4 has a by-product in the form of an assembly of five pentagons into one. The small sketch on the left of Fig. 5.6 shows how five strip elements fit together. The assembly is symmetric, but not minimal.

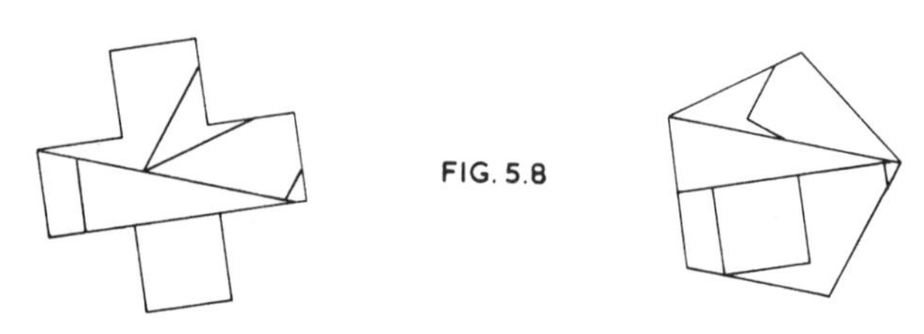

FIG. 5.8

A superposition tends to be uneconomical if the angle between strips is small, since too many cuts are likely to cross in an elongated common area. An exception is the *TT*2 dissection of Greek cross and pentagon in Fig. 5.7. It requires seven pieces, whereas other superpositions all seem to require more. Actually the number of pieces in the common area is quite large, namely 14; the saving grace is that only half of them are needed for the dissection. Figure 5.7 may look rather bewildering, so I make the concession of showing the dissected polygons in Fig. 5.8.

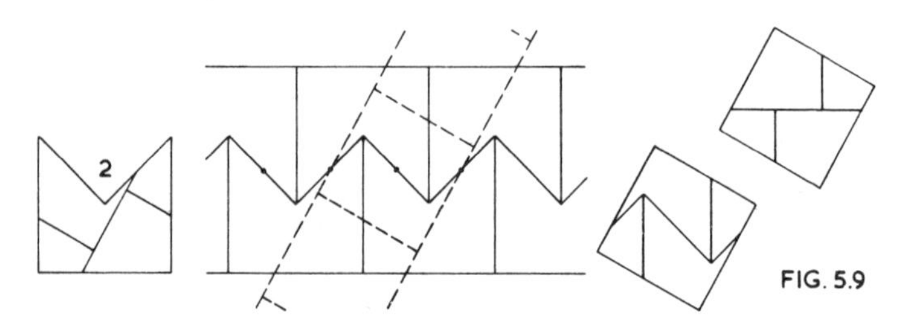

FIG. 5.9

The strip element in Fig. 5.5 does not extend across the strip. Another one like this is shown in Fig. 5.9, in which the element is a miter (or, if you like, a stool end). The problem of finding a four-piece dissection of miter and square seems to be insoluble. But as the figure shows, two miters can be dissected by identical cuts into two squares, with only four pieces per miter. (It happens quite often that a dissection of two polygons into two is more economical than that of one into one.) This dissection, which is not a *TT*2 but a *PT*, illustrates the point that in a *TT*2 dissection a transverse line in each strip must pass through the center of the common area. Here a transverse line in the strip of squares does not. Compare Fig. 5.2.

It was noticed by Dudeney that *TT* dissections of trapezoids or triangles can be made in the form of hinged models. Half

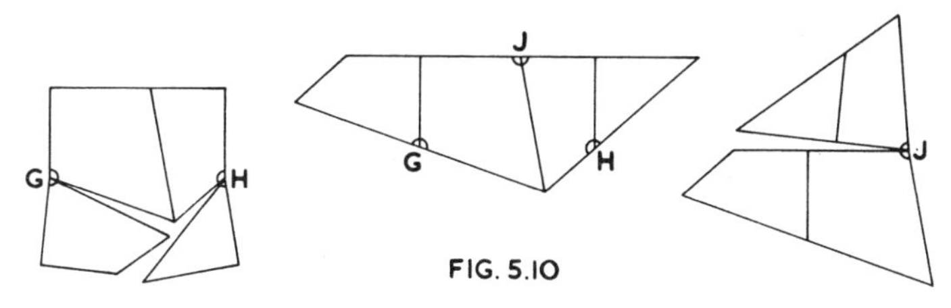

FIG. 5.10

of the common area in Fig. 5.2 is reproduced in the middle of Fig. 5.10, and the four pieces are hinged at the corners *G, H, J,* to form an open chain. The first and last pieces, swung downwards about *G* and *H,* give the square as on the left, and the two right-hand pieces, swung together upwards about *J,* give the triangle as on the right. It is similar with any *TT*2 dissection. And you will find that in a *TT*1 dissection such as Fig. 5.1, hinges at *A, D, E* enable each strip element to be transformed into the other. We have already seen in Fig. 1.10 that a *Q*-slide can be made in the form of a hinged model, the pieces forming a closed chain. It will be found later that dissections obtained by other methods also can be made in the form of hinged models.

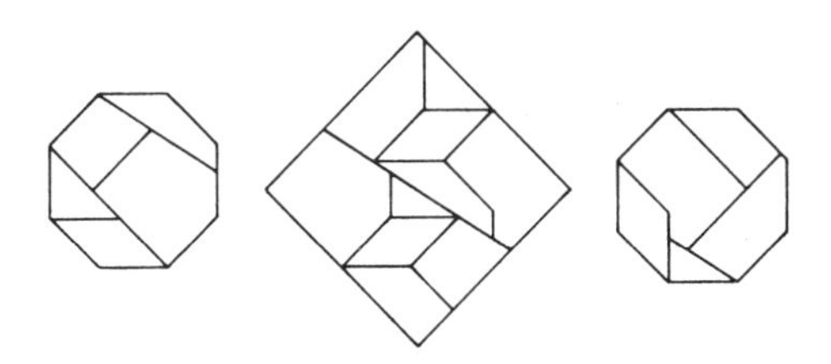

6

PIECEMEAL DISSECTIONS

Sometimes there is a way of changing a polygon into two strip elements, which is easier than changing it into only one. Thus a heptagon is easily divided into a triangle and two trapezoids, and the latter can be combined into a parallelogram as in Fig. 2.8. On the other hand it is relatively troublesome to change it into a single strip element. (This problem is taken up in Chapter 7.)

Suppose now we want to dissect the heptagon into a square, whose side is to be of length s. We can do so by dissecting both the parallelogram and triangle into rectangles of length s and stacking the rectangles. Their widths of necessity add up to s, so we get a square.

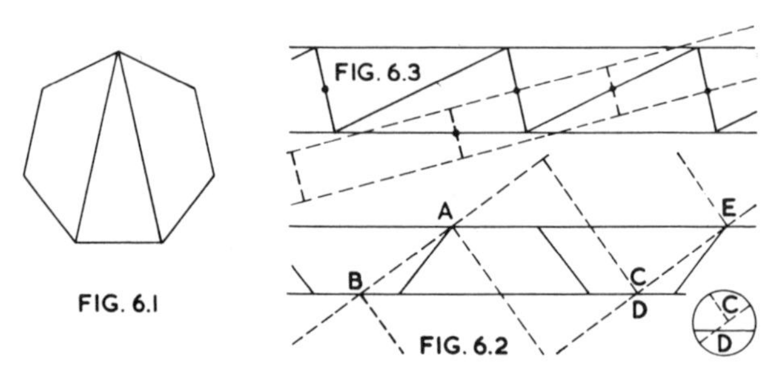

FIG. 6.1 FIG. 6.2 FIG. 6.3

The parallelogram is dissected by PP, in a manner similar to Dudeney's hexagon-square dissection, as is seen on comparing Figs. 3.2 and 6.2. We get a five-piece dissection by the skin of our teeth. In Fig. 6.2 the vertices A and B of the rectangle are placed on the edges of the other strip, and a drawing would have to be unpractically large and accurate to show beyond doubt that C then lies above the horizontal line, as is preferable, and not below. So we must resort to calculation. The details will not be given here, but it can be found with the help of Appendix C that

$$\angle ABC = 35° \ 36', \angle ABD = 35° \ 51' > \angle ABC.$$

It follows that C is just above the horizontal line (and, therefore, so is E).

FIG. 6.4

The triangle is dissected by $TT2$. Figure 6.3 shows the dissection, which is an application of the method described in Chapter 5. The combination of Figs. 6.2 and 6.3, giving the complete dissection of heptagon and square, is shown in Fig. 6.4.

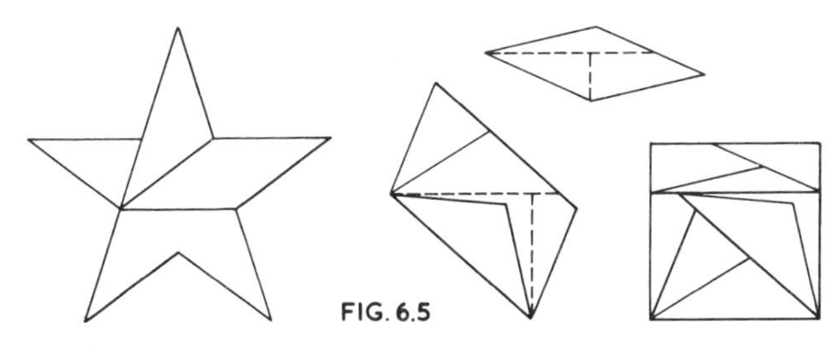

FIG. 6.5

Another piecemeal dissection is that of pentagram and square. The former is changed as in Fig. 6.5 into two parallelograms. Since these are changed into rectangles by easy *PP* dissections, drawings of superposed strips are dispensed with, and all the steps are squeezed into one figure.

The name *piecemeal dissection* for one of this kind, although apposite enough, is not really suggestive of its nature. But I cannot think of a better one.

7

STRIPS FROM TESSELLATIONS I

The polygons so far considered are easily made into strip elements; polygons such as the heptagon, nine-gon, and dodecagon are a different matter. But there are methodical ways of getting strip elements from them too, by means of tessellations.

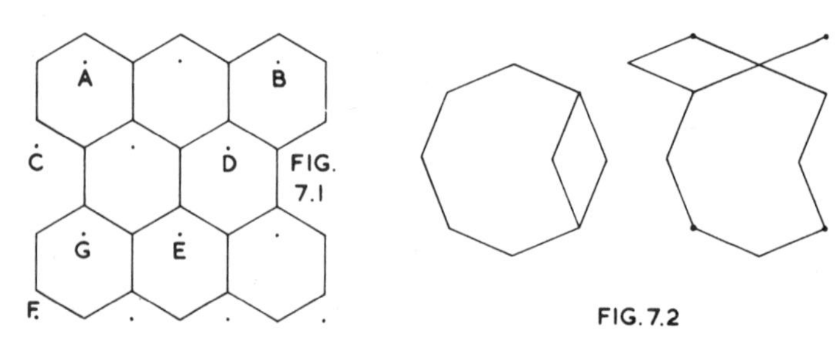

FIG. 7.1

FIG. 7.2

All the points marked by dots in Fig. 7.1 are congruent. They lie in parallel straight rows, and lines through parallel rows of a set of congruent points will be called *congruent lines.* Thus lines through *AB, CD* are congruent, so are lines through *AD, CE,* so are lines through *CF, AG.* The part of the tessellation bounded by two congruent lines (at a suitable spacing) is a strip such as we can use for finding dissections. Thus, the strips in Figs. 2.8 and 2.9 are given by *CF, AG* and *AB, CD,* and the strip in Fig. 4.5 is given by *AD, CE.* The first and second of these three are *P*-strips and the third is a *T*-strip, so we see that a tessellation may give strips of either kind. If the congruent lines bounding a *P*-strip in Fig. 7.1 are displaced without changing their spacing or direction, they continue to bound a *P*-strip.

The full lines in Fig. B15 * constitute a tessellation of Greek crosses. You can satisfy yourself that all the Greek-cross strips

* Figures B1-B16 are all in Appendix B.

illustrated so far (Figs. 2.1, 2.18-2.21, 4.6) are obtained by drawing congruent lines on either this tessellation or its mirror image. Here again, an infinite number of *P*-strips is obtained from any one by continuous displacement.

The hexagon and the Greek cross are, of course, already tessellation elements. Other polygons that are not are sometimes easily made so. Thus, an octagon can be cut as in Fig. 7.2 into two pieces, and rearranging these gives the element of the tessellation in Fig. B8. If the strip element in Fig. 2.10 by chance escaped your notice, vertical congruent lines on Fig. B8 would find it for you.

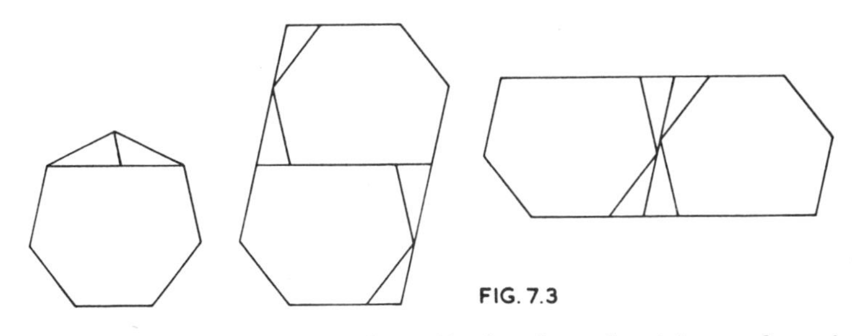

FIG. 7.3

The foregoing strips could easily be found without dragging in tessellations; they are only simple examples to show the procedure. But to discern a strip element directly from a drawing of a heptagon or dodecagon requires more imagination than most of us have, and these are the kinds of polygon for which a tessellation is really useful. The first step is to make the polygon a tessellation element; one simple way for a heptagon is shown in Fig. 7.3. A pair of the elements can be assembled in two ways, also shown in Fig. 7.3, to make a hexagon with opposite sides parallel. And hexagons like this can be assembled into a tessellation, the one obtained from the first hexagon being shown in Fig. B7. Both *P*- and *T*-strips can be obtained from this tessellation, even though half of the elements are inverted, by drawing parallel lines through the centers of symmetry marked. It will be found that bisecting any *T*-strip gives a *P*-strip. The restriction to lines through centers of symmetry, when half the elements of a tessellation are inverted, is analogous to the restriction in the use of *T*-strips.

FIG. 7.4

FIG. 7.5

FIG. 7.6

FIG. 7.7

The dodecagon plus two small triangles is the element of the well-known tessellation shown in Fig. 7.4. This tessellation has an application described in Chapter 12, but otherwise it is not much use if we want to dissect the dodecagon alone, seeing that it includes two unwanted triangles. For this purpose we must make a tessellation element of the dodecagon on its own, by cutting it into a few pieces and rearranging, and the difficulty about this is the abundance of 150° angles and the dearth of 210° angles for them to nestle into. But Fig. 7.4 shows that if one of the pieces is a small triangle, we can fit three portions of perimeter round it and, thus, find a home for six of the 150° angles. Experiment on this basis led to cutting the dodecagon as in Fig. 7.5. The pieces thus obtained can be arranged into a tessellation element in at least two ways, as in Figs. 7.6 and 7.7. The tessellations appear in Figs. B12 and B13.

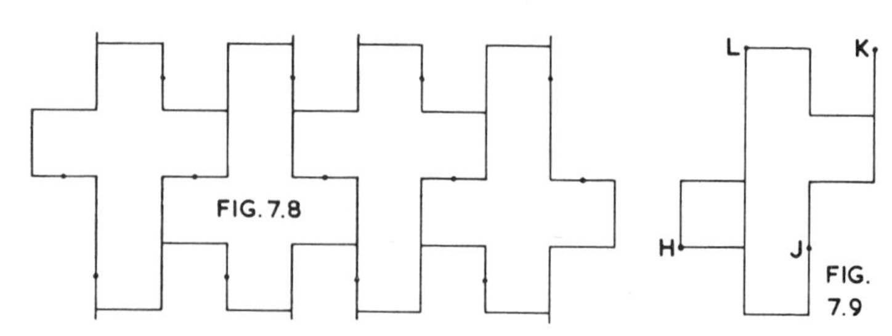

FIG. 7.8

FIG. 7.9

It is convenient to have a term for a parallelogram such as *HJKL* in Figs. 7.6 and 7.9. There is no need to invent one, for the theory of elliptic functions provides us with the term *period parallelogram.* The same theory provided us with the term *congruent.*

The dodecagon tessellations just found have different period parallelograms. Because of this a greater variety of strip elements derived from the one polygon is available. A further reason for seeking more than one tessellation will emerge in Chapter 9. It is even worth while to examine tessellations that are not the most economical; that is, the polygon is cut up into more than the minimum number of pieces that will make it a tessellation element. For instance a Latin cross is already a tessellation element, as shown by Fig. 7.8; but the less economical element in Fig. 7.9 (two pieces instead of one) will be found later to have a most excellent application. It is in any case natural that the second Latin-cross tessellation, in which all elements are the same way up, should be more useful, for Fig. 7.8 yields strips only if the parallel lines are drawn through centers of symmetry; whereas in Fig. 7.9 they need not pass through *HJ* and *LK* or *HL* and *JK,* but can be parallel thereto.

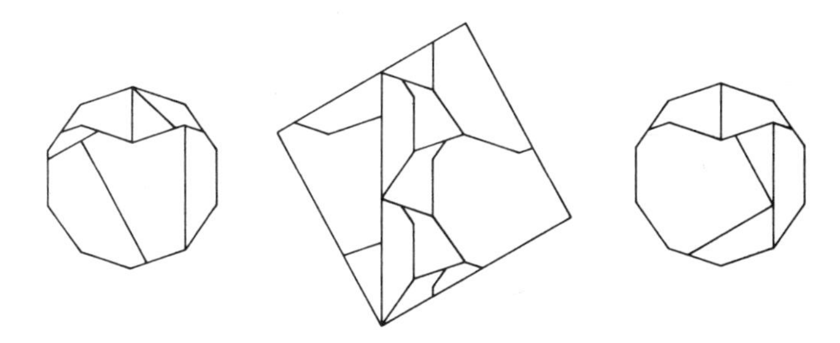

8

TRIANGLES AND QUADRILATERALS

In this chapter we consider the dissection of any triangle or quadrilateral into any other triangle or quadrilateral. The shapes of the two polygons will be taken to be not too disparate, so that the most favorable but still general dissection will be possible. The extension to less favorable cases causes no difficulty.

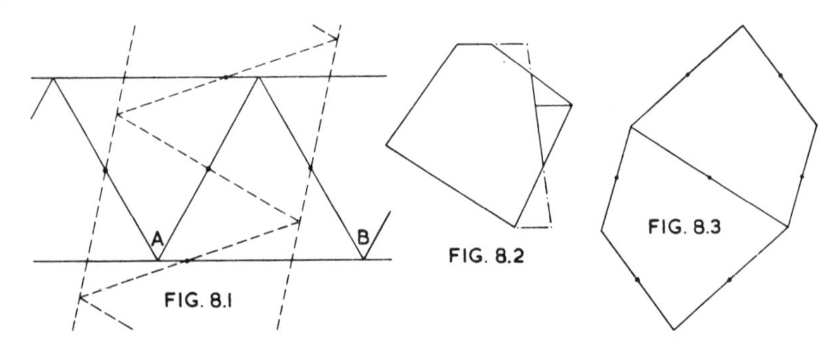

FIG. 8.1 FIG. 8.2 FIG. 8.3

Any two triangles can be dissected by *TT*2. Unless a triangle is equilateral, three different *T*-strips can be formed from it. (This applies to isosceles triangles too, the third strip being obtained by turning over one of the others.) Of the nine combinations of strips from two triangles, the best is likely to be that in which *AB* in Fig. 8.1 is equal to or slightly greater than the width of the other strip. The general dissection requires four or more pieces, and clearly this applies also to dissections of a trapezoid and a triangle, or of two trapezoids.

Any quadrilateral also can be dissected into another one or into a triangle by *TT*2. Every quadrilateral (even a re-entrant one) is a tessellation element, for two of them can always be paired to form a hexagon with opposite sides parallel, as in Fig. 8.3. This polygon, as we saw in Fig. 7.3, is a tessellation element. Several T-strips can be obtained by drawing parallel lines through centers of symmetry, which here are midpoints of sides.

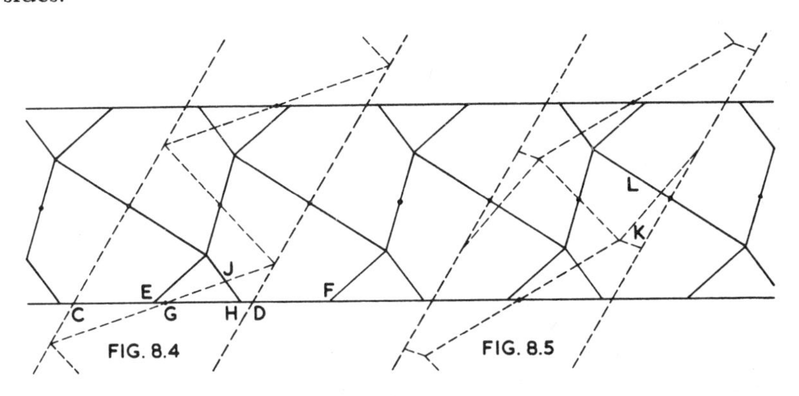

FIG. 8.4 FIG. 8.5

The dissection of quadrilateral and triangle in Fig. 8.4 requires six pieces—two more than for two triangles. One of the extra pieces is due to the fact that CD and EF do not in general coincide, nor, in consequence, do their midpoints G and H. If H is on the right of G as shown, then two cuts cross as at J; if H is on the left of G, so that E is on the left of C, then there is a crossing involving an edge of the sloping strip.

For the same reason two further cuts are required in dissecting two quadrilaterals, making eight pieces in all. Referring to Fig. 8.5, we see that the extra piece K is cut by line L.

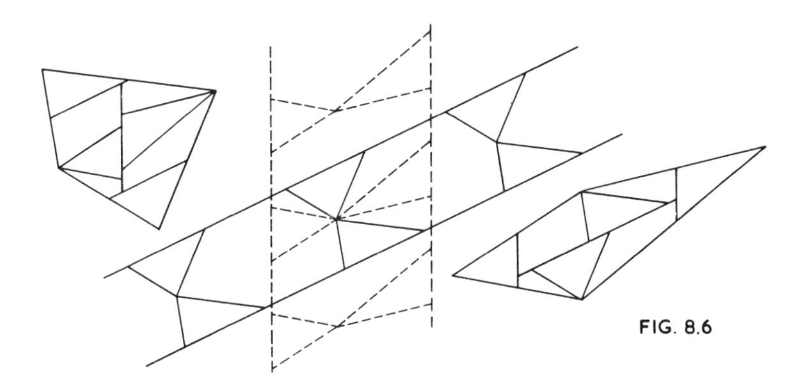
FIG. 8.6

The same sort of dissection is obtained if a quadrilateral is replaced by a pentagon with two sides parallel, which also is a

tessellation element, as can be seen from Fig. 7.3. An otherwise intractable irregular pentagon can always be changed to one with two sides parallel by cuts as in Fig. 8.2. This then would be the first step in dissecting it.

Dissections involving triangles, it seems, are minimal only if one uses *TT*2. But this does not always apply to the dissection of two quadrilaterals. Each can be cut into three pieces to make a *P*-strip element, with a good chance that a *PP* superposition will give a minimal dissection. Figure 8.6, due to W. H. Macaulay, shows one. The rearrangements are included, to show how they (as well as the common area) have some degree of symmetry—as much as can be expected when the quadrilaterals are irregular.

It is almost certain that the numbers of pieces (four, six, eight) for general dissections of two triangles, quadrilateral and triangle, and two quadrilaterals are truly minimal. Even today a competent and sufficiently determined mathematician could probably prove this with full rigor. But rigorous proofs of most conjectures like this lie far in the future. Can a regular pentagon, for instance, be dissected into a square in fewer than six pieces? At present one can only challenge a doubter to do it in five, and this state of affairs will persist for a long time to come.

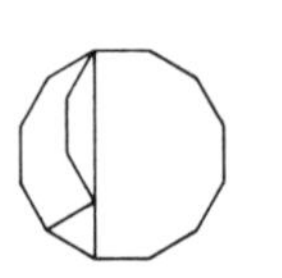
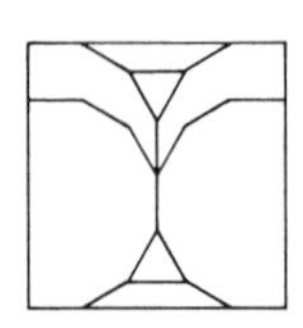
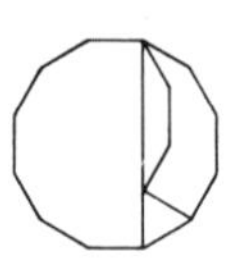

9

DISSECTIONS FROM TESSELLATIONS I

The Greek cross has already been dissected here into various polygons, but not yet into a square. Minimal dissections could be found by superposing a strip of squares on strips made up of the elements in Figs. 2.18-2.20, and a further, nonminimal dissection could be found from a strip made up of the element in Fig. 2.21. In the first three the strips cross at right angles, so transverse lines in the strip of squares can be made coincident with the edges of the Greek-cross strip. But strips are not the most natural or effective device to use for this dissection. We can dispense with them and use instead the tessellations they were derived from.

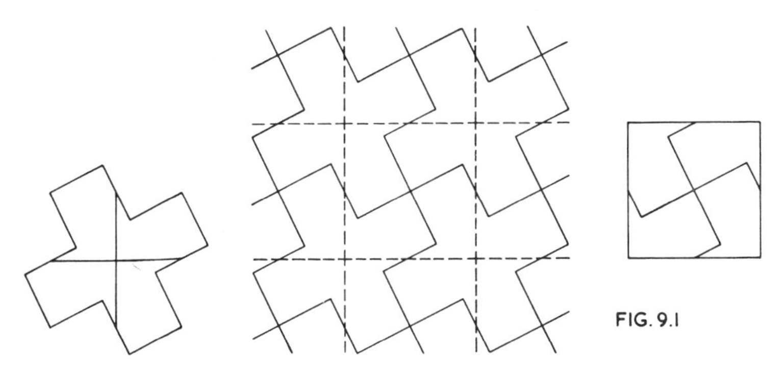

FIG. 9.1

The period parallelogram of the Greek-cross tessellation is a square. And one gets a dissection of cross into square merely by superposing the tessellations of cross and square, making points congruent in the one coincident with points congruent in the other. In Fig. 9.1 congruent points in the tessellation of crosses, such as the centers of the crosses, coincide with congruent points in the tessellation of squares, namely their vertices. If we move the tessellation of squares in any way, keeping the lines in it horizontal and vertical, we always have a dissection of cross and square. Some other polygons whose tessellations have a square as period parallelogram are shown in Figs. 9.2-9.5, a set of four

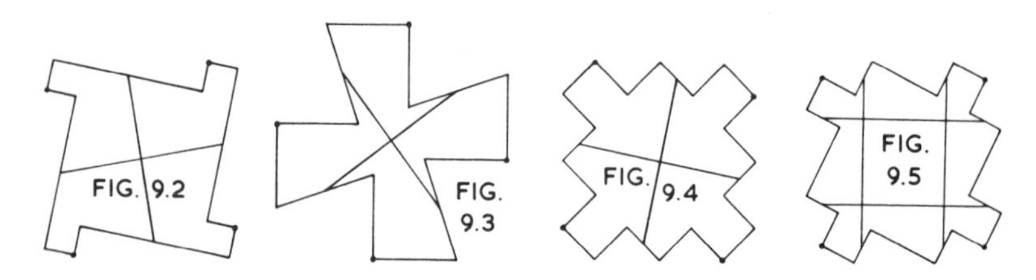

congruent points in each being marked by dots. (You are recommended to sketch the tessellations on graph paper.) The cuts show the symmetric dissections of the first three into one square, and of the fourth into four squares, all easily found by superposing tessellations. Other dissections are found by shifting one of the tessellations. A superposed tessellation of Greek crosses will also give four-piece symmetric dissections.*

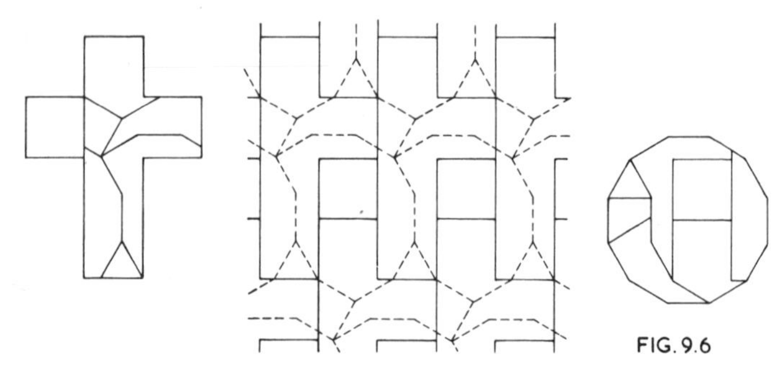
FIG. 9.6

The same kind of superposition is possible with any two tessellations that have the same period parallelogram, and in which all elements are the same way up. An example in which the period parallelogram is not a square is the dissection of Latin cross and dodecagon. It can be verified that the parallelograms

* There is a four-piece dissection of Fig. 9.5 into four Greek crosses. A Nobel prize will not be awarded for discovering it.

HJKL in Figs. 7.6 and 7.9 are identical. Superposing the tessellations leads to a dissection of cross and dodecagon as in Fig. 9.6.

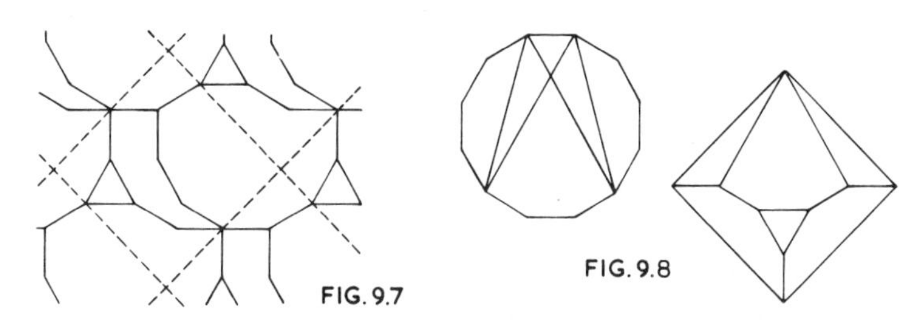

FIG. 9.7

FIG. 9.8

The period parallelogram of the tessellation with element as in Fig. 7.7 is a square. A six-piece dissection of dodecagon and square can certainly be obtained by superposing the tessellations; Fig. 9.7 shows one. But the tessellation has found out for us that the chord subtending four sides of a dodecagon is equal to the side of the equivalent square. Knowing this, we can find several dissections merely by trial on the dodecagon, and in almost embarrassing abundance; Figs. 9.8-9.10 show three of them. The second is probably the neatest, but the first is unusual in that the cuts in the dodecagon are so easy—just four chords.

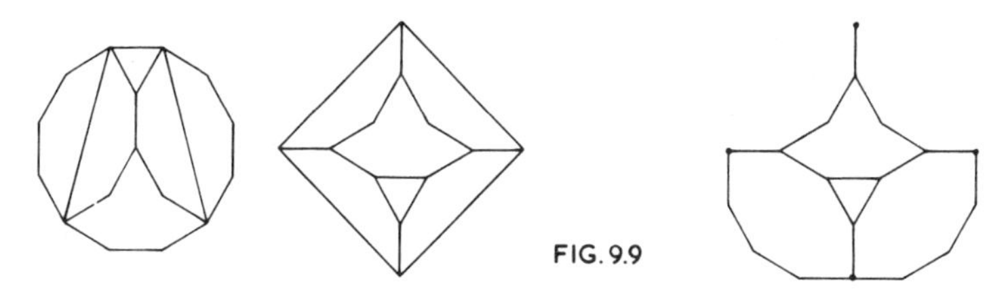

FIG. 9.9

In their turn, these dissections lead to ways of making the dodecagon a tessellation element that we might otherwise have missed. We can then see if anything can be done with the further tessellations thus found. For instance, the tessellation with element as on the right of Fig. 9.9 gives the Greek-cross dissection in Fig. 9.11. The possibility of new and useful strips also should not be overlooked; Fig. 9.10 includes a strip element that will be used later.

The Maltese cross and Nazi swastika are not tessellation elements, but are easily made so, as in Figs. 9.12 and 9.13. Two

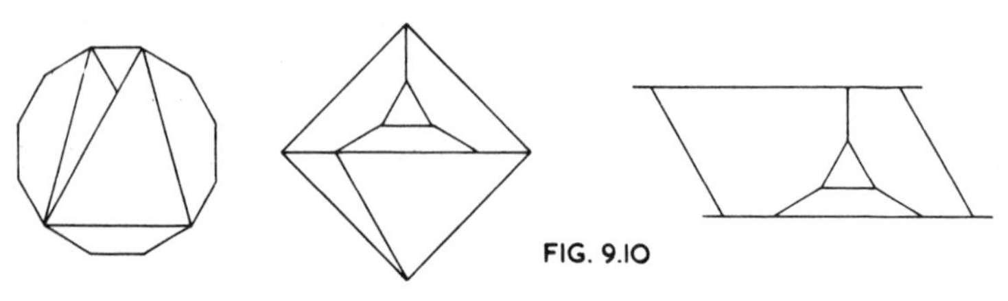

FIG. 9.10

different tessellations are possible in each case, and it is gratifying to find that the period parallelograms of one of them are squares. The square tessellations are shown combined in Fig. B16. From them we find minimal dissections of swastika into square as in Fig. 9.14, and of Maltese cross into Greek cross.

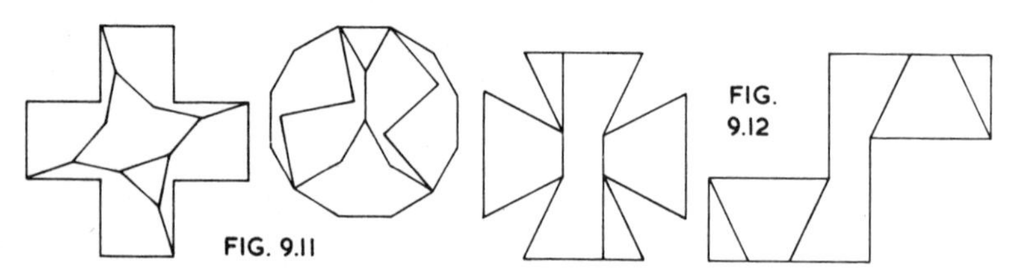

FIG. 9.11

FIG. 9.12

An eight-piece dissection of Maltese cross and square can be obtained by superposition from the tessellation of Fig. 9.12. It is not illustrated, because it is not minimal. The Maltese cross can be dissected into a swastika merely by superposing the tessellation elements in Figs. 9.12 and 9.13.

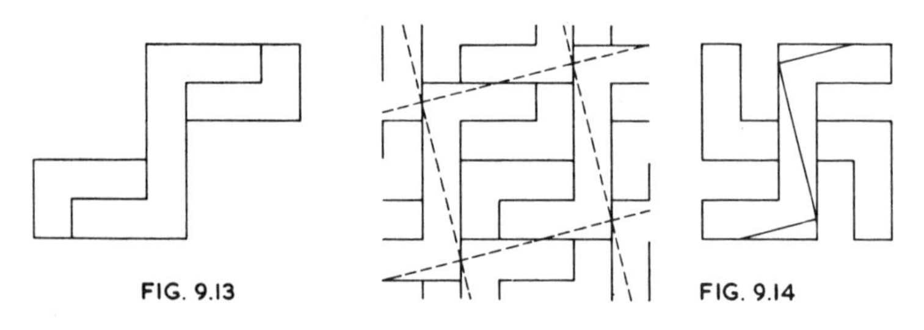

FIG. 9.13

FIG. 9.14

There is another form of swastika, shown in Fig. 9.15. Unlike the Nazi form, it is a tessellation element as it stands, and because of this the dissections into square and into Greek cross require fewer pieces. Figure 9.16 provides an alternative dissection of Nazi swastika and Greek cross. The preliminary step is to cut

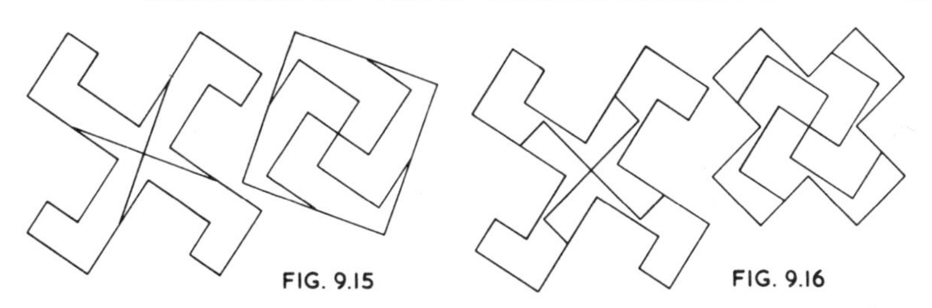

FIG. 9.15

FIG. 9.16

off the 2×1 rectangles forming the hooks of the swastika and displace them so as to get the other form.

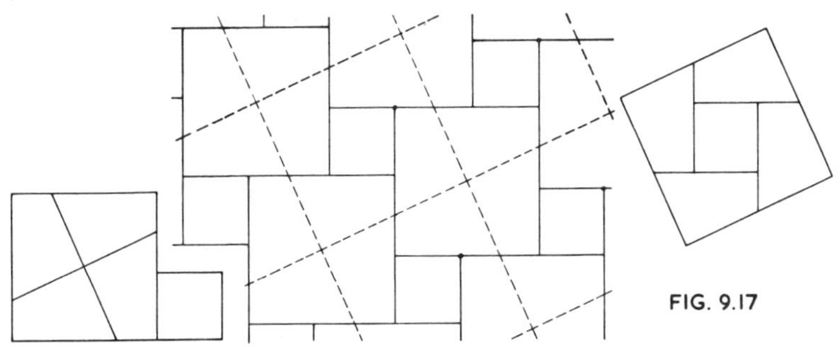

FIG. 9.17

A simple tessellation that will find many uses later is that of two squares, equal or unequal. Figure 9.17 shows that its period parallelogram is a square, whence we get dissections of two squares into one. The figure shows the pieces for Henry Perigal's well-known visual demonstration of Pythagoras's theorem. If the two squares are combined into a single polygon, one of its dissections into a square has only three pieces. (Join the dots.)

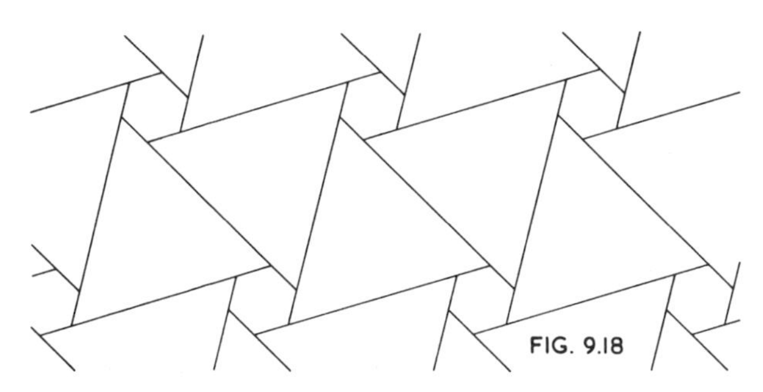

FIG. 9.18

The tessellation of Fig. 9.17 can be obtained from an ordinary tessellation of squares by mutually displacing the members of each set of four adjacent squares. A similar displacement in a tessellation of triangles or hexagons gives tessellations whose element consists of two triangles plus a hexagon. The tessellations in Figs. 9.18 and 9.20 look different, but each merges into the other via the semiregular tessellation $(3{\cdot}6)^2$ in Fig. 9.19. The period parallelograms are similar to those of the original tessellations. This makes Figs. 9.18 and 9.20 useful, as we shall see in Chapters 12, 20, and 21.

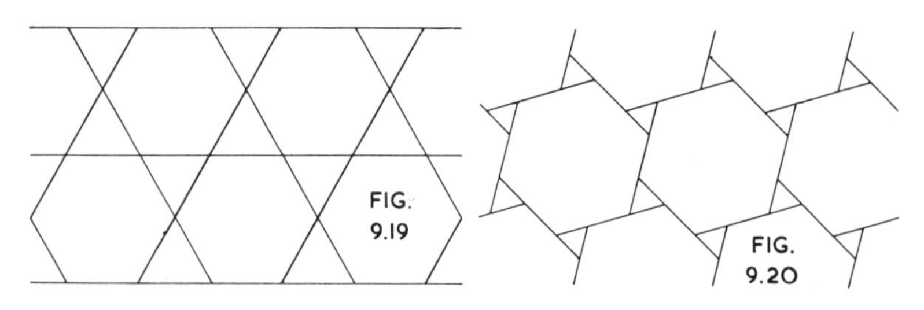
FIG. 9.19 FIG. 9.20

A tessellation may have a period parallelogram, but half of its elements may be inverted, as in Fig. 7.8. If two of this kind have the same period parallelogram, they can be superposed to give a dissection. There is a restriction, analogous to that in the use of *T*-strips, that centers of symmetry (marked by dots in Fig. 7.8) must coincide; otherwise, one gets a dissection of two polygons into two, but not of one into one. The only such dissection in this book is the one referred to in connection with Fig. 18.11.

Occasionally two strips have the same width, and a dissection can be obtained with the strips coincident. It will be realized that this is in effect a superposition of tessellations. Figures 4.12 and 23.1 are examples.

10

CLASSIFICATION

Dissections obtained by superposing tessellations will be called *T*-dissections. They complete a tentative scheme of classification for dissections of rectilinear polygons by rectilinear cuts, which is as follows:

P, Parallelogram slide.
Q, Quadrilateral slide.
R, Rational. Includes step dissections.
S, From strips. Subdivided into *PP, PT, TT*1, and *TT*2.
T, From tessellations.
U, Unclassified.

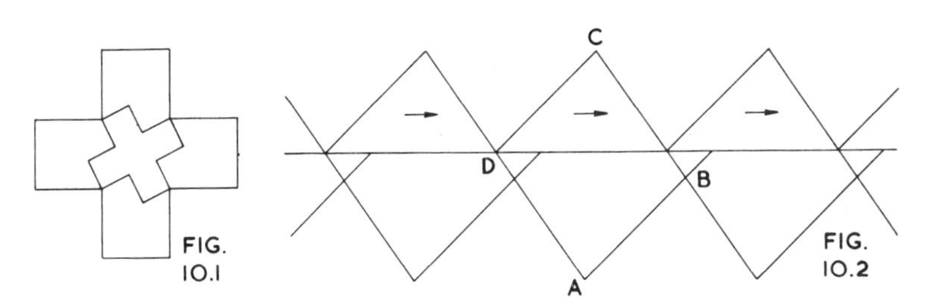

FIG. 10.1

FIG. 10.2

You will understand that I *had* to call the third class rational, because this word begins with an R. *T*-dissections could be subdivided on lines similar to the subdivision of *S*-dissections, but with the present paucity of examples this would be classification for its own sake.

U-dissections comprise those that cannot be worked out by any of the methods described in this book, so that the would-be solver must rely entirely on his own ingenuity. One example is Dudeney's Greek-cross puzzle in Figs. 10.1 and 10.3. Most of

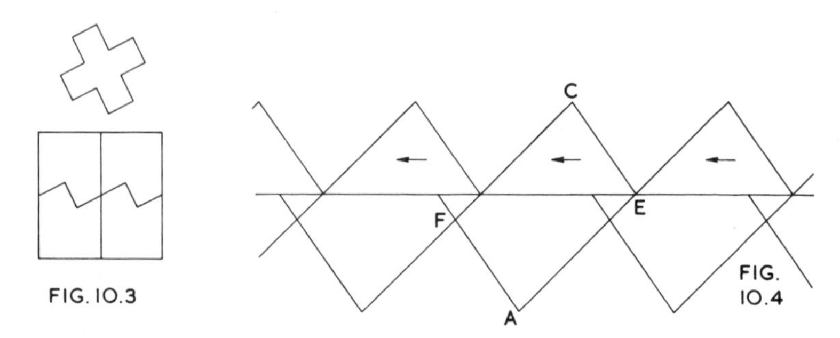

FIG. 10.3

FIG. 10.4

the dissections in Chapter 20 are also *U*-dissections. Perhaps someone will discover a new method of working out dissections that will reduce the *U*'s and increase the non-*U*'s.

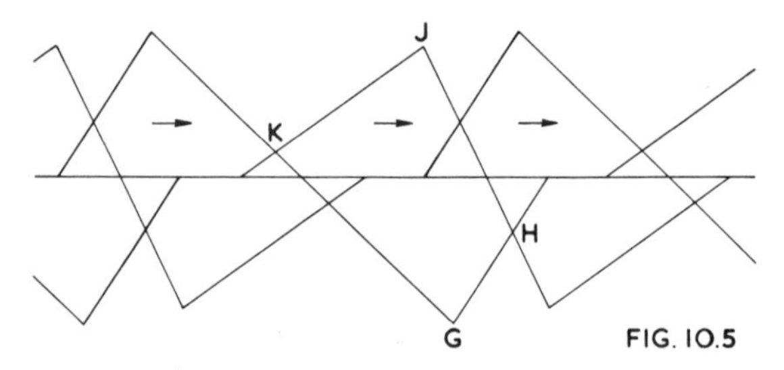

FIG. 10.5

You may think the term "*Q*-slide" inapposite, for where is the slide? Figures 10.2, 10.4, 10.5, and 10.6 supply the answer. The first two, relating to a *P*-slide, show how a chain of parallelograms *ABCD* is dissected, by sliding the parts above the horizontal line, into a chain of parallelograms *AECF*.

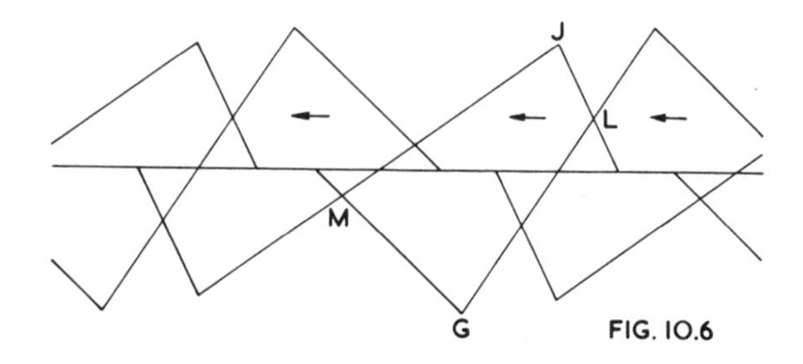

FIG. 10.6

Figures 10.5 and 10.6 show how a chain of quadrilaterals *GHJK* (alternate ones inverted) is similarly dissected into a chain of quadrilaterals *GLJM*. Here is the *Q*-slide.

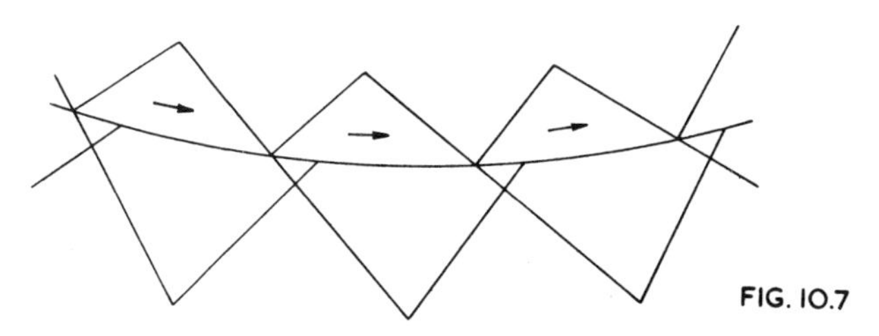

FIG. 10.7

This way of presenting the slides suggests a further possibility; what if the shear-line is a circular arc? Such a slide exists too, and is shown in Figs. 10.7 and 10.8. In this circular slide not

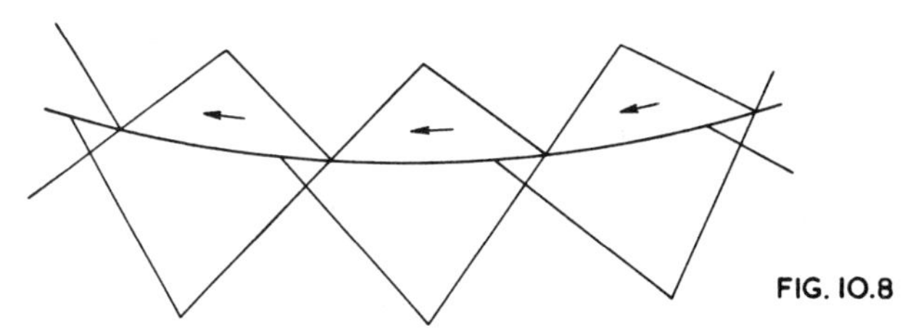

FIG. 10.8

only are the sides changed, but so are two of the angles, namely, those near the shear-line. One is increased, and the other reduced, by an amount equal to the angular displacement of the parts above the shear-line. The relations between the dimensions of the two quadrilaterals are rather intricate, which helps to explain why this brainchild appears to be stillborn. (But Fig. 1.17 is similar.)

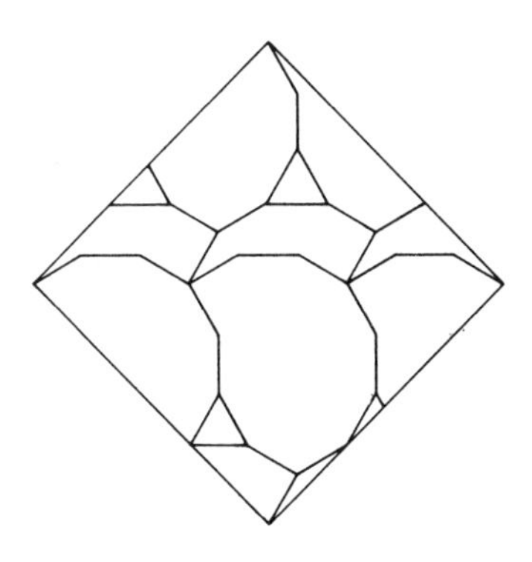

11

DISSECTIONS FROM TESSELLATIONS II

In some tessellations the elements fall naturally into sets of two, such as the set in Fig. 7.8. In others they fall into sets of more than two. A simple example is one whose element is a gable end with either a full roof, or an undersized one as in Fig. 11.1. Here we have sets of four that form, according to taste, either a distorted cross or a square with a triangular excrescence near each vertex. If the centers of symmetry of two tessellations of this kind have identical patterns, a dissection of one polygon into one is obtained by superposing with centers of symmetry coincident. Figure 11.1 shows how a dissection of gable end and square is thus obtained. It can be seen that if the lines crossing at A rotate round A keeping the polygon in one piece, the latter is always the element of a tessellation having the same centers of symmetry.

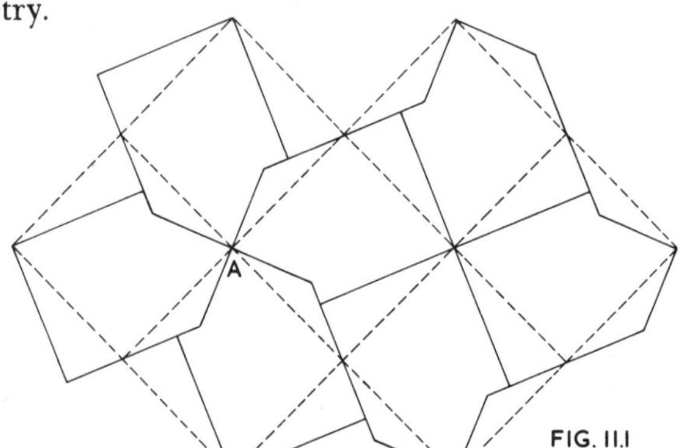

A

FIG. 11.1

A tessellation of hexagons can be regarded as made up of sets of three hexagons, each set consisting of three that meet at a point. Each set is a *compound* element, let us say. In Fig. 11.2 the small hexagons can be associated in sets of three, forming compound elements with centers *B, C, D, E.* It is easily seen that these centers have the same pattern as centers of single hexagons that are three times as big. So we can superpose a tessellation of the larger hexagons, as shown, and get an assembly of three hexagons into one. The centers of compound elements and of single hexagons are shown coincident. This is not necessary, for one tessellation can be translated without restriction relatively to the other. But here coincident centers happen to give a minimal dissection.

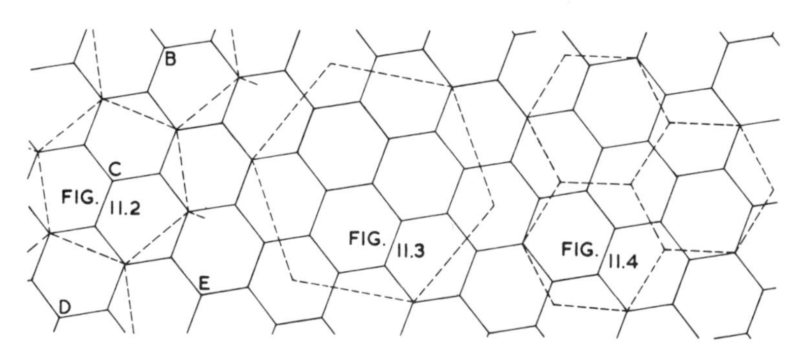

The same applies to sets of seven hexagons, each set consisting of six hexagons grouped round a central one. Because of this you can superpose the tessellations and find assemblies of seven hexagons into one as in Fig. 11.3, and of seven into three as in Fig. 11.4. In the same way you can handle dissections of 13, 19, 21, 31, • • • hexagons, if you want to. There is a fuller discussion in Chapter 21, Section 21.2.

A dissection that belongs to this chapter is the one in Fig. 11.5, for the second figure shows an intermediate step giving an element that is one of a set of four. This is a case in which superposing tessellations is quite unnecessary; anyone clever enough to think of the intermediate arrangement would not need tessellations to go further and find the square. But the methods using strips and tessellations have the two purposes of being heuristic and of providing a system of classification.

The dissection in Fig. 11.5 was discovered by A. E. Hill. It is

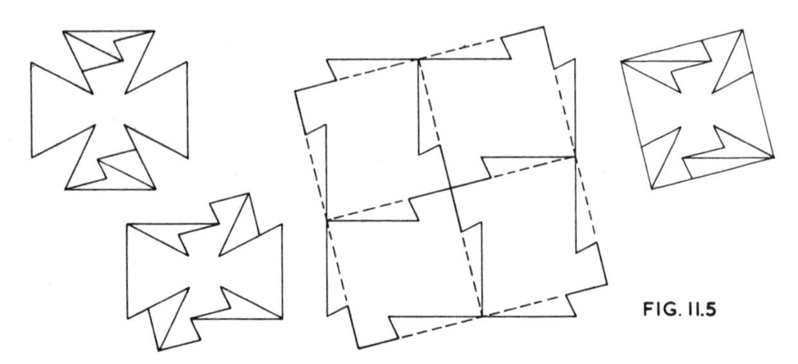
FIG. 11.5

surely the most remarkable one there is. I would like to have discovered it myself.

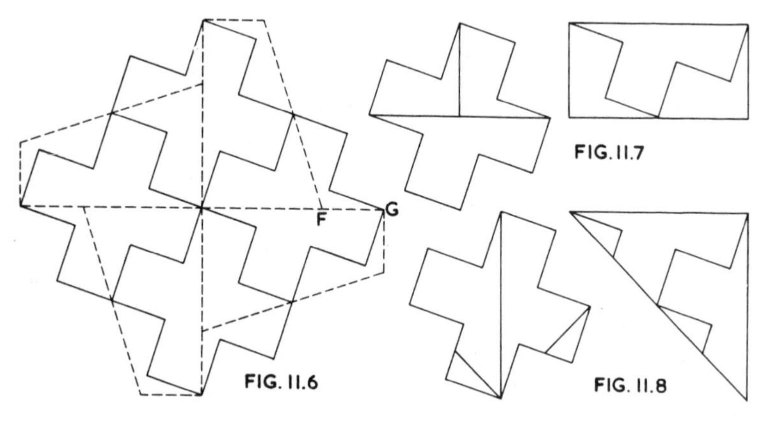

FIG. 11.6

FIG. 11.7

FIG. 11.8

If a square is bisected by any line through its center, then four of the half-squares thus obtained form a compound element of a tessellation, shown by broken lines in Fig. 11.6. (The bisector need not be a single straight line, but must be symmetric about the center of the square.) We can also regard each set of four Greek crosses that meet at a point as a compound element; its shape is that of Fig. 9.5. And the centers of symmetry of the two tessellations have the same square pattern. Superposing them gives a dissection of Greek cross and half-square that has three pieces or four when the bisector is a straight line, the number depending on how near F is to G. If the half-square is a rectangle, we have a three-piece dissection; if it is an isosceles triangle, a four-piece one. These two cases, both by Dudeney, are illustrated.

12

COMPLETING THE TESSELLATION

An octagon by itself is not a tessellation element, but with the addition of a small square it is. The period parallelogram of the tessellation, which is shown in Fig. 12.1, is a square. Looking back at Fig. 9.17, we see that the period parallelogram of a tessellation having as element a large square plus a small one is also a square. Let the area of the large square be equal to that of the octagon, and let the side of the small square be equal to that of the octagon. Then superposing the tessellations gives the dissection

octagon + small square →
square equivalent to octagon + same small square.

Take away the small square from both sides of this, and you get James Travers's dissection of octagon and square.

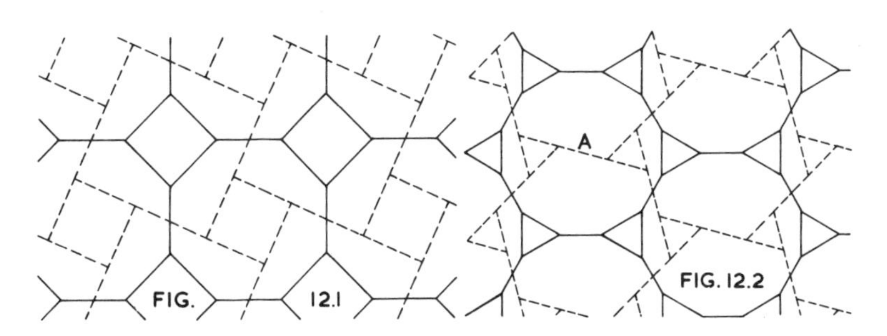

FIG. 12.1 FIG. 12.2

The preface to Dudeney's *Puzzles and Curious Problems* contains the statement, "It is remarkable that the regular octagon can be cut into as few as four pieces to form the corresponding square." But this I think is an error. The five-piece dissection in Fig. 12.1 is one of those which we can be morally certain are best possible.

Similar is the dissection of dodecagon and hexagon. The dodecagon plus two small triangles is a tessellation element (see Fig. 7.4), and, as we saw in Fig. 9.20, so is a hexagon plus the same two small triangles. Again the tessellations have the same period parallelogram, so we superpose and get Fig. 12.2. A line such as A appears to coincide with a diameter of the dodecagon, but actually it is at an angle of about 0° 32.6′.

This dissection was first discovered by Irving L. Freese, a Los Angeles architect who died in 1957. One of his minor efforts has already appeared in Fig. 5.6. The present dissection is a masterpiece, as are the two in Figs. 21.27 and 21.28.

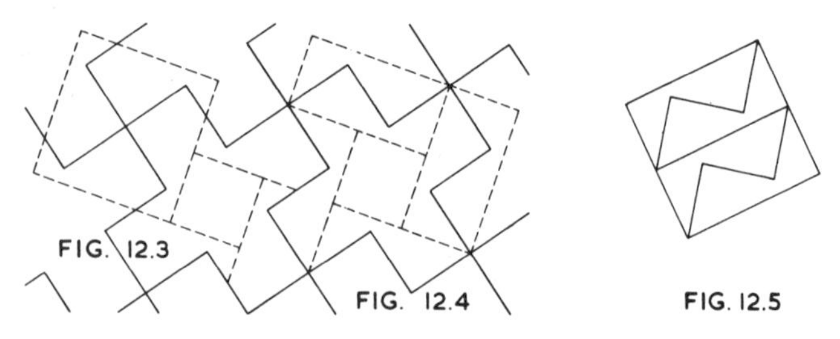

FIG. 12.3 FIG. 12.4 FIG. 12.5

Occasionally a polygon for dissection has a piece cut out of its interior. If the external and internal boundaries of the polygon are favorable, we can complete the tessellation by restoring the piece cut out. This applies to the dissection into a square of a Greek cross with a square cut out. We complete the Greek-cross tessellation by filling up the vacant squares, and use it in combination with Fig. 9.17. Dudeney's four-piece dissection in Fig. 12.3 requires that the cut-out square in the cross have a particular angle of tilt; otherwise more pieces are required. But its location and size can be varied appreciably without making more than four pieces necessary. The dissection in Figs. 12.4 and 12.5, also by Dudeney, is applicable when the square removed is central and has one fifth of the area of the cross.

A different example of completing the tessellation is the dissection of a square, mutilated by cutting off a corner, into a whole square. Four of the mutilated squares, grouped as in Fig. 12.6

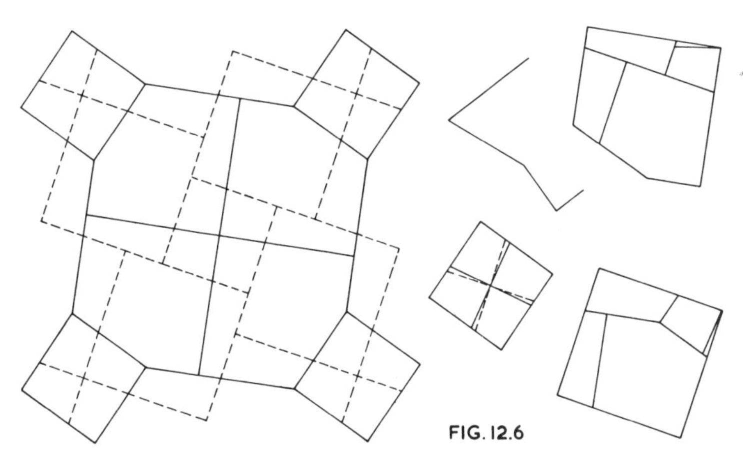

FIG. 12.6

into an irregular octagon, combine with a small square to form the compound element of a tessellation somewhat similar to the octagon-square tessellation in Fig. 12.1. And four whole squares, grouped into a large square, combine with the same small square to form a two-square tessellation, again similar to the one in Fig. 12.1. The tessellations have the same period parallelogram, so they can be superposed. The superposition must have centers of symmetry coincident, for we want a dissection of one polygon into one, not of four into four. It differs from the previous ones in that the added squares also are cut, and the quarter-squares they are cut into are of different shapes in the respective tessellations. Therefore, the one quarter-square must be dissected into the other. The necessary cut is found by superposing the small squares complete with cuts, and we get a five-piece dissection.

In similar fashion a square with an added triangular spout can be dissected into a square.

13

STRIPS FROM TESSELLATIONS II

The number of *P*-strips obtainable from a tessellation can be embarrassingly large. Thus, any pair of horizontal lines at the right spacing, drawn on Fig. 7.1, give a hexagon strip. And sometimes one feels that a dissection that has been found could be improved, if only the right choice of strip among an infinite number of them could be made. A way of coping with this difficulty will now be explained with reference to a *PT* dissection of Greek cross and hexagon.

If we get the hexagon strip by drawing horizontal lines on Fig. 7.1, then something we do know about it is its width. So if it is superposed on the Greek-cross *T*-strip with element as in Fig. 4.6, we know the angle between the strips, whence the hexagon strip can only be in one of the positions shown in Fig. 13.1. The second position is better because it has fewer crossing points, so we confine our attention to it.

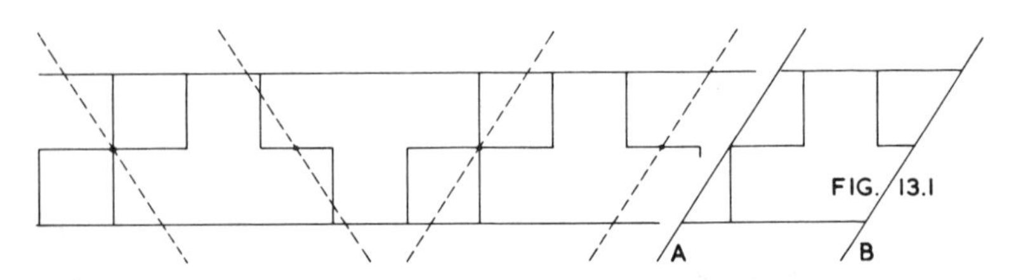

FIG. 13.1

We now know which part of the Greek-cross strip will coincide with the common area; it is the parallelogram on the right of Fig. 13.1. And the elongated sides *A, B* thereof are edges of the hexagon strip, that is, they are congruent lines in the hexagon tessellation. So we superpose this parallelogram on the hexagon tessellation of Fig. 7.1 with *A* and *B* horizontal. (It is helpful to make *A* and *B* fairly long, lest one unwittingly rotate them.)

We find that the superposition with fewest cuts in the common area is the one in Fig. 13.2, and the best hexagon strip has thus been found.

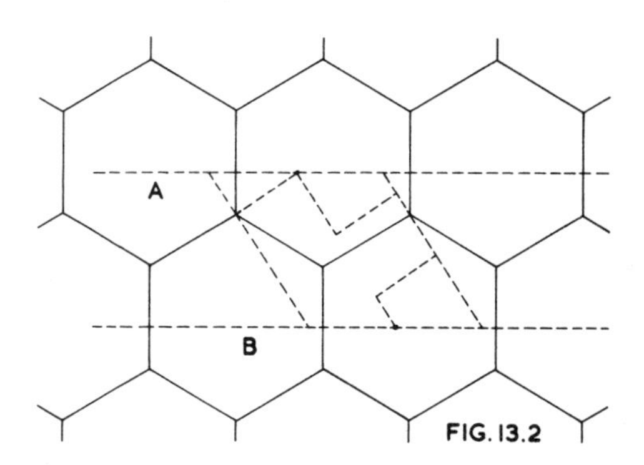

FIG. 13.2

This procedure works well with *PT* dissections such as Fig. 13.2, in which the part of the *T*-strip that coincides with the common area is easily located. Sometimes one can locate it in a *P*-strip too, as in the next example.

Assembling three Greek crosses into a single square was easy enough to be discussed in Chapter 1. But assembling them into a single Greek cross is not so easy. The difficulty arises in part from the number of one-cross strips that can be used. Even if attention is confined to strips parallel to *CD* in Fig. 13.3, the number of them is infinite, so it is desirable to proceed as in Figs. 13.1 and 13.2.

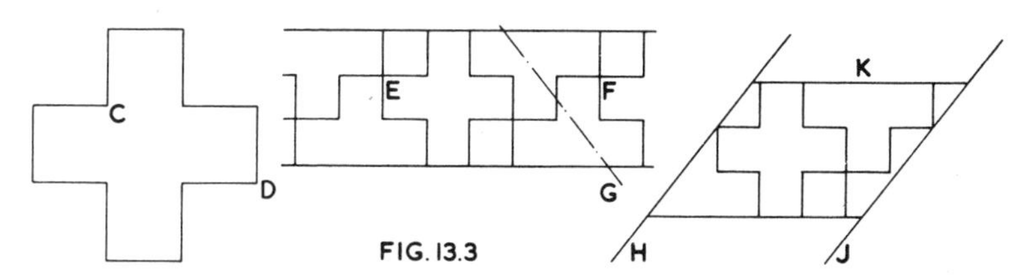

FIG. 13.3

On the other hand only one three-cross strip seems worth trying, namely that of Fig. 2.1. For experiment on a tessellation will show that the only other strips that need be taken seriously are those of width *EF*, and these cannot be used, being wider than *CD*.

As before, we know the slope of the one-cross strip when superposed on the three-cross one; it is that of line *G* or *H*. The next question is, is there any way of placing the edges of the one-cross strip on the three-cross one that is most promising? There is, for if and only if they slope as at *H* and *J* and pass through *E* and *F*, then each edge crosses cuts in the three-cross strip at only one point, namely *E* or *F*. We have thus chosen a parallelogram *K* in the three-cross strip to coincide with the common area.

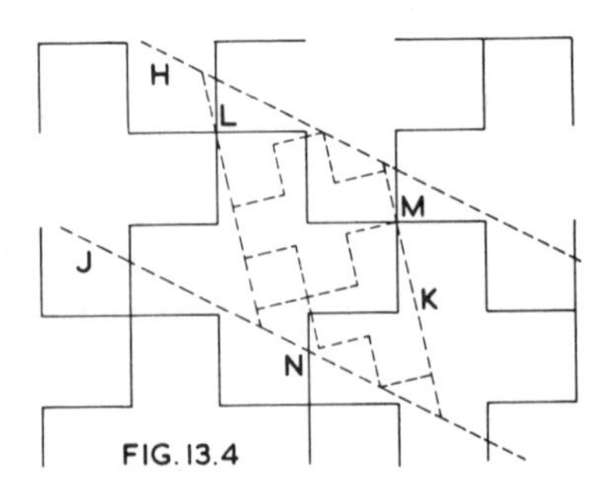

FIG. 13.4

The best one-cross strip can now be found by superposing *K* on a one-cross tessellation, as in Fig. 13.4, with lines *H* and *J* congruent. The common area must have inside it at least two broken lines that cross the zigzag lines *LM* and *MN*, and its roughly vertical sides must cross other lines at least once as at *L* and *M;* we conclude that the dissection illustrated cannot be improved on.

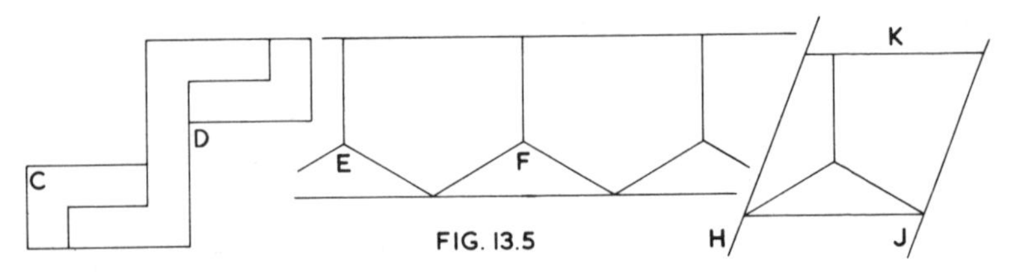

FIG. 13.5

Another example of this procedure is the dissection of swastika and hexagon, using the square tessellation in Fig. B16 and the strip with element as in Fig. 2.9. Thus, in this case, the simpler polygon (the hexagon) takes the place of the more complex one (the three crosses) in the previous case.

The lettering in Fig. 13.5 corresponds to that in Fig. 13.3, to show how the procedure is the same. As before, there are two possible slopes for lines *H* and *J*, but the two corresponding parallelograms *K* are mirror images, so it suffices to use one of them and turn it over. Also, we pass *H* and *J* through vertices of the hexagon, this being the most natural thing to do. The best swastika strip is the one obtained by superposing as in Fig. 13.6, in which it will be seen that parallelogram *K* has been turned over.

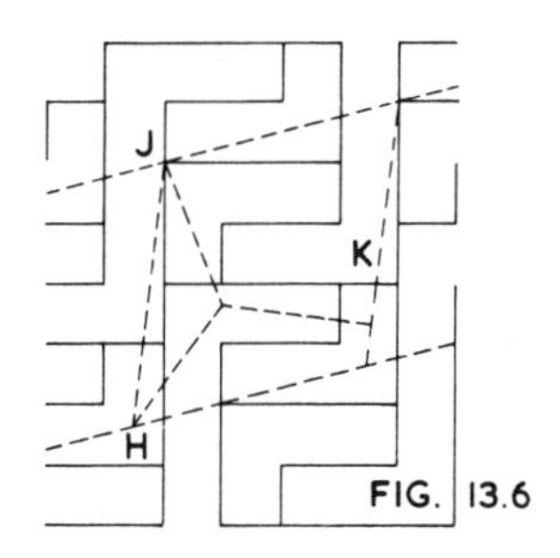

FIG. 13.6

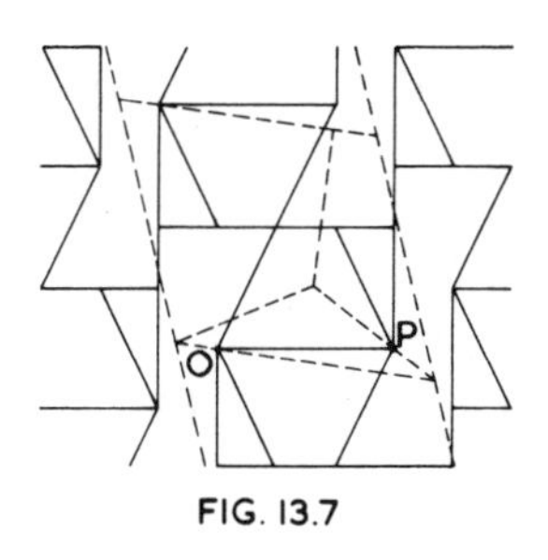

FIG. 13.7

Getting a dissection of swastika and hexagon emboldens us to try dissecting the Maltese cross, which has a similar tessellation, into a hexagon. A 14-piece dissection is shown in Fig. 13.7. The hexagon element has been superposed so that lines in it pass through *O* and *P* in the cross tessellation.

In the last two dissections one has the choice of fixing a common area beforehand in the hexagon strip or in a strip derived from the tessellation of swastika or cross. But where on earth can one find a likely strip in the swastika tessellation, let us say? It has lines all over the place, and all possible strips derived from it look equally unpromising. So one fixes a common area in the hexagon strip instead.

14

RATIONAL DISSECTIONS

A census of published dissections reveals that rational ones are the most numerous. No doubt this is because they are fairly easy to devise, and puzzles based on them can be made neither too easy nor too hard. The census also reveals that in nearly every case the problem is to dissect some polygon into a square. This specialization appears to be justified, in that extensions to figures other than the square tend to be profitless. So in this chapter we confine ourselves to rational dissections into a square.

FIG. 14.1

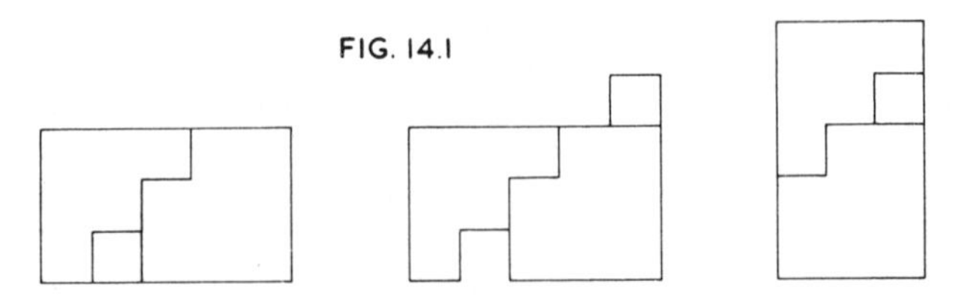

Figure 1.15 shows that a step dissection of rectangle and square does not require a rectangle in which length/width is a perfect square of the form $(n + 1)^2 / n^2$, where n is an integer. True enough the rectangle must be one in which length/width is a perfect square, and only if it is the square of a quotient of consecutive integers do we have a two-piece dissection. In describing step dissections with more pieces, we shall not consider the dissection of a rectangle with dimensions $m^2 \times n^2$ into a square of side mn, but the dissection of an $m \times n$ rectangle into an $n \times m$ one with no rotation of pieces. This makes the rise and tread of the steps uniformly 1, which is convenient for experiment on graph paper. The dissection into a square is retrieved by multiplying lengths by m and widths by n.

There is a three-piece dissection of any $m \times n$ rectangle, if

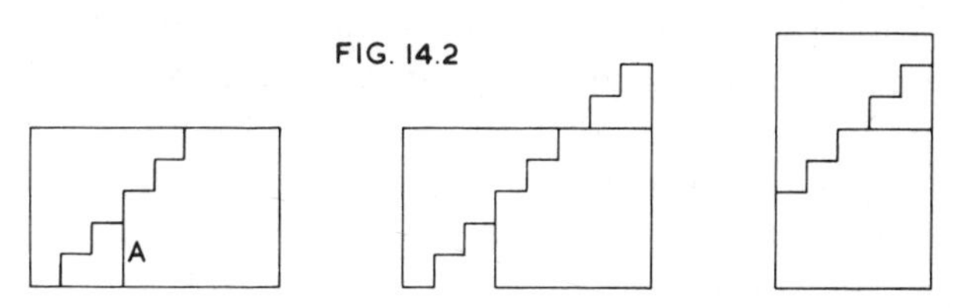

FIG. 14.2

not too long and narrow, into an $n \times m$ one. Figures 14.1 and 14.2 show the dissections with $m = 5$, $n = 3$ and $m = 8$, $n = 5$. As is clear from the intermediate step, this is a bumpy kind of *P*-slide.

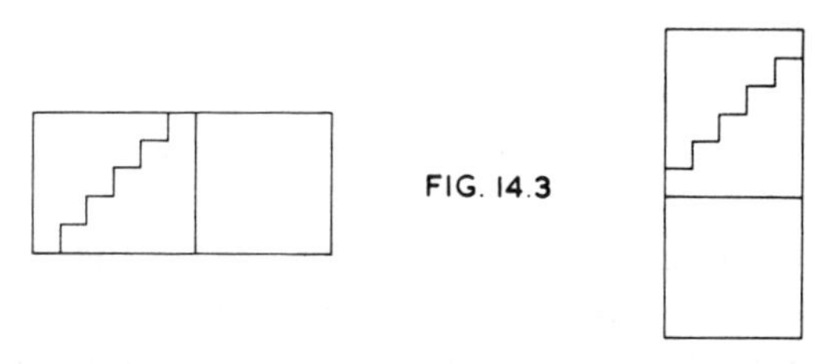
FIG. 14.3

Most dissections like this are no better than a plain *P*-slide unless a puzzle is restricted to step dissections, but there is one exception. This is when the vertical cut *A* in Fig. 14.2 has been moved to its extreme right-hand position, as in Fig. 14.3. Since n can be any integer instead of 5 as shown, there is a step dissection of a $(2n + 1) \times n$ rectangle into an $n \times (2n + 1)$ one.

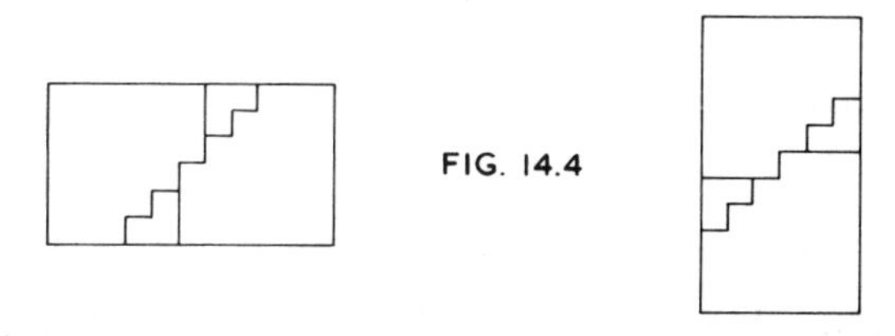
FIG. 14.4

This corresponds to the dissection of a $(2n + 1)^2 \times n^2$ rectangle into a square. Now in a plain *P*-slide a rectangle with a three-piece dissection into a square cannot have length/width greater than 4. So the step dissection enables us to exceed this limit. The maximum value of length/width is 9 (when $n = 1$), giving a rather trivial dissection; the next largest value is 25/4 (when $n = 2$), giving Fig. 1.15.

If there are step dissections corresponding to the *P*-slide, why not others corresponding the the *Q*-slide? They exist, as Fig. 14.4 shows, but the trouble is that the sides of the polygon have to be parallel to either the rise or the tread of the steps, and this ties us to parallelograms or combinations of them. Thus, the *Q*-slide has no advantage, apart from symmetry, to compensate for the extra piece. As regards symmetry, it will be noticed that in Figs. 14.1-14.4 the cuts in the initial and final rectangles are mirror images.

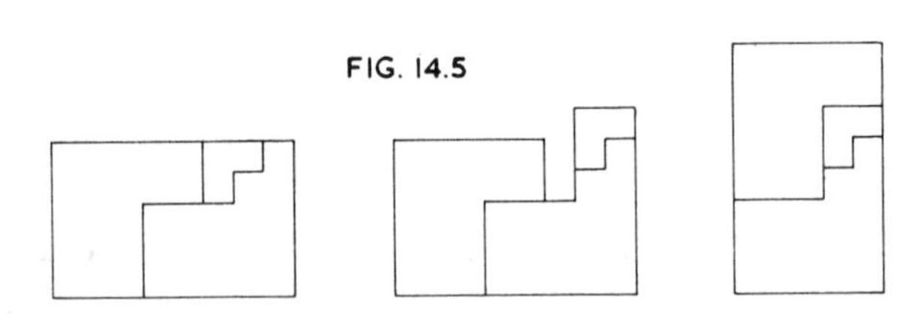

FIG. 14.5

It might be thought that all steps must have the same rise and tread. Figure 14.5 shows the contrary. The intermediate stage consists in preparing for the step movement by dissecting the 3×2 rectangle in the upper right corner into a 2×3 one. The 3 and 2 can be replaced by other integers, the preliminary step dissections need not be on the top step, there can be combinations of them, and so on.

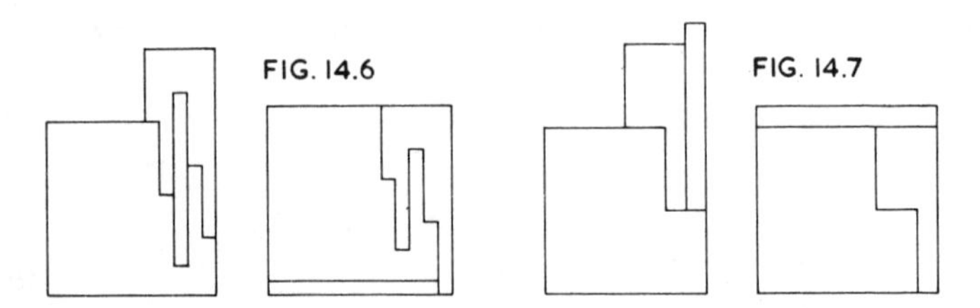

FIG. 14.6 FIG. 14.7

The *dramatis personae* in Loyd's and Dudeney's puzzles spent a lot of time assembling square patchwork quilts or the like, cutting along seams. Figures 14.6 and 14.7 show two of their more *recherché* efforts, these being visual demonstrations of the numerical identities

$$12^2 + 5^2 = 13^2, \; 8^2 + 4^2 + 1^2 = 9^2.$$

They are clearly minimal.

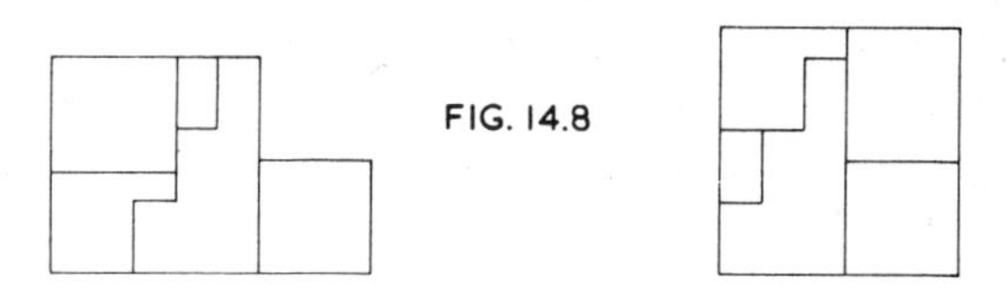
FIG. 14.8

There is an endless variety of such dissections. Considering only the case of assembling two squares into one, denote the sides of the squares by a, b, c (no common factor, $a < b$, $a^2 + b^2 = c^2$). Then a four-piece dissection requires $b = c - 1$, as in Figs. 1.19 and 1.20. If this condition is not satisfied, we can often get a five-piece dissection by changing one of the small squares into a gnomon which fits round two sides of the other small square. Thus, if $a = 8$, $b = 15$, $c = 17$, Fig. 14.8 shows how you get the gnomon. This step dissection of two squares into one has five pieces like a *T*-dissection based on Fig. 9.17, and so is minimal. It can be verified that the condition for a minimal step dissection is:

If $(c + a)/b$ or $(c + b)/a = p/q$ (this is now irreducible), then $p \leqslant 2q + 1$.

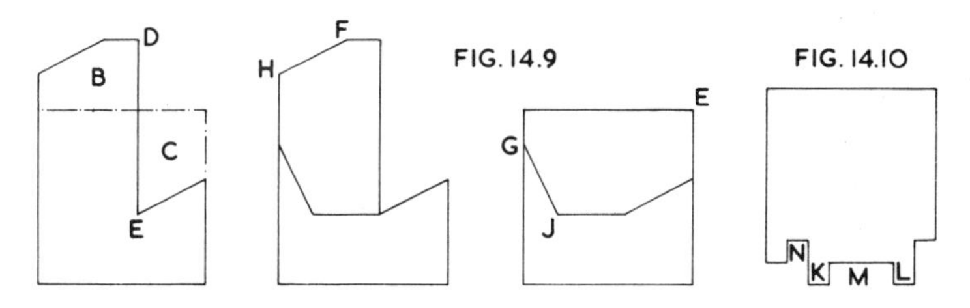

FIG. 14.9 FIG. 14.10

The only other rational dissection that will be touched on is the two-piece one with a quarter turn. An example was illustrated in Fig. 1.18. It is possible to solve this kind line by line, as will be done with Loyd's puzzle in Fig. 14.9. Merely cutting along the chain-dotted line is no good, for piece *B* does not fit into space *C*. But as *DE* is equal to the side of the equivalent square, we can seek some piece, bounded on the right by *DE*, which can be given a quarter turn anticlockwise to make *DE* the top of the square. Comparison of the initial figure and the square shows that F determines G, H determines J, and so on round to vertex E of the square.

Examining Fig. 14.10 (Fig. 1.18 without the cut) in the same way, we see that both K and L have to be moved, so both must be in the rotated piece. Anticlockwise rotation is clearly no good because of the space M. But clockwise rotation to bring L over N will lead without fail to the cut in Fig. 1.18.

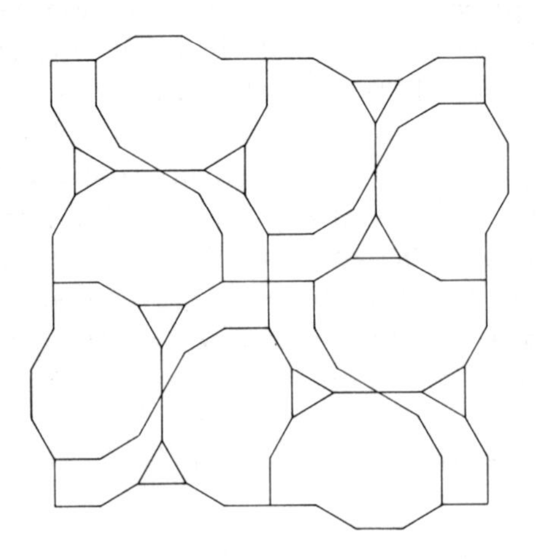

15

THE HEPTAGON

Only one of the heptagon dissections to be described uses a strip that is not obtained from the tessellation in Fig. B7. Even so, the exception is obtained from its element. Cutting and rearranging the element as on the left of Fig. 15.1 makes it a *T*-strip element, and the *TT*2 dissection into a triangle is on the same lines as nearly all the other triangle dissections in this book. The very small piece is a blemish in this dissection; but no other strip gives a nine-piece one.

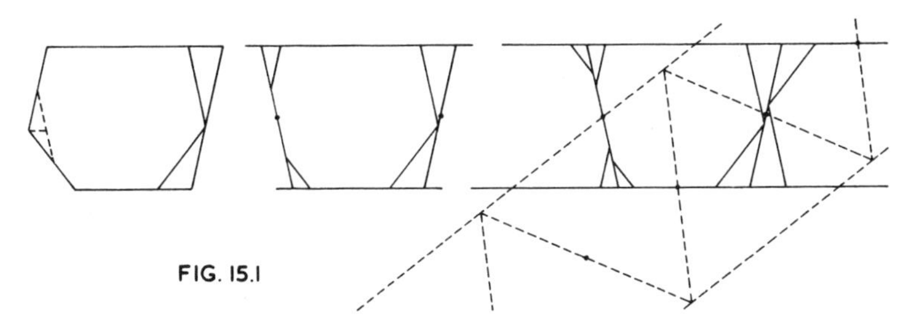

FIG. 15.1

Minimal dissections into other polygons use strips obtained from the tessellation, sometimes avoiding a piece as small as in Fig. 15.1. (Small pieces cannot, however, be avoided altogether—the heptagon, although beautiful, is reluctant.) To make clear how the strips are obtained from the tessellation, the strip elements in Figs. 15.2-15.6 are drawn with their edges at the same angles as they would be in Fig. B7.

Two *T*-strip elements are drawn in Fig. 15.2, because a single element does not reach from edge to edge of the strip. The *P*-strip with element as in Fig. 15.3 is obtained by bisecting the *T*-strip. In Fig. 15.4, we again have the element of a *T*-strip which, when bisected, gives a *P*-strip.

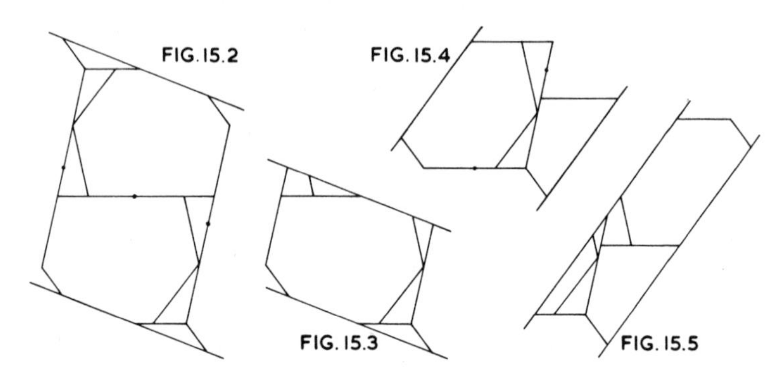

The edges of the T-strip element on the left of Fig. 15.6 are vertical in the tessellation. This is easily proved, for A and B are the midpoints of CD and EF, and since $GD = HF$, AB is parallel to the line through the midpoints of CG and EH, which is vertical.

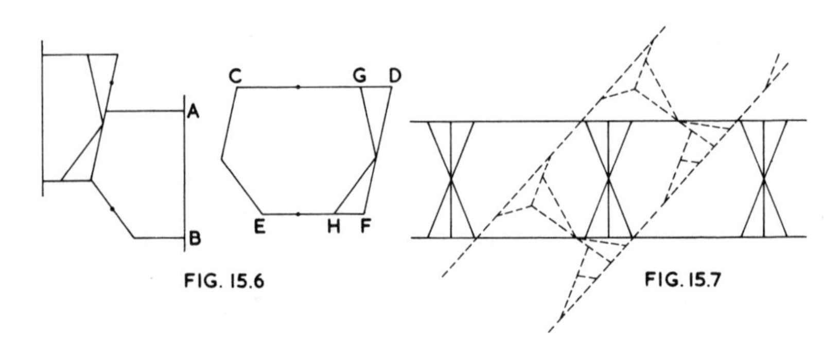

Using three of the strips illustrated, we find dissections of heptagon into octagon, pentagon, hexagon, hexagram, and Latin cross. They are shown, in that order, in Figs. 15.7-15.11.

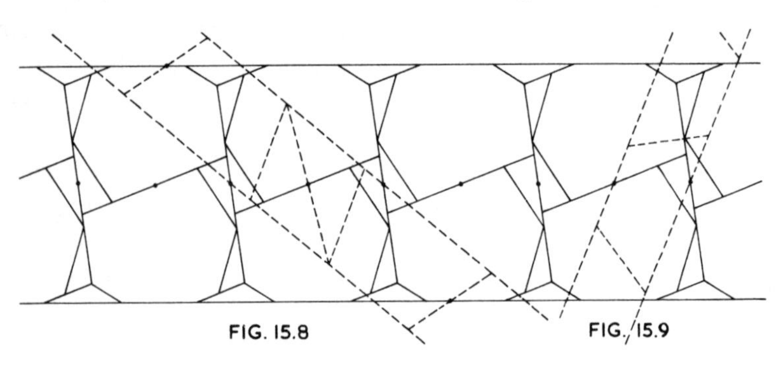

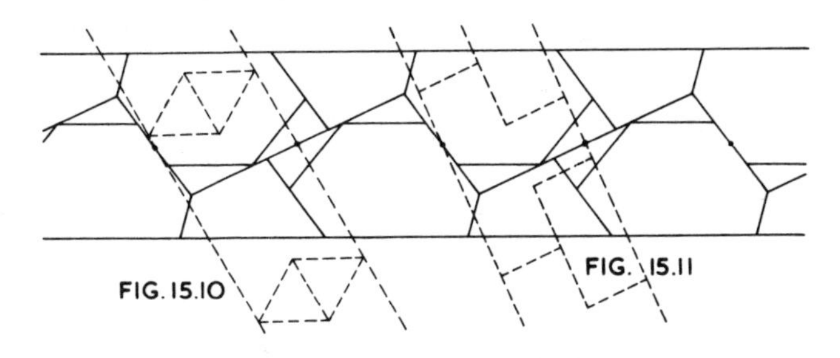

It will be noticed that all the heptagon dissections described are based on only one of the two tessellations mentioned in connection with Fig. 7.3. A few sporadic trials give the impression that the other tessellation is inferior, in that dissections based on it require as many pieces or more, and one of the pieces is very small indeed. But there was no systematic investigation, so it is quite possible that in one or two cases the other tessellation gives a better result.

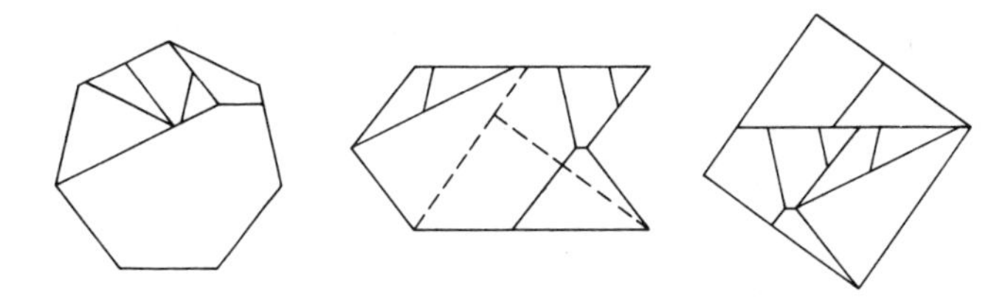

16

THE NINE-GON

One might expect the nine-gon to be more amenable to dissection than the heptagon, because angles between sides and/or chords have sensible values like 60°. But experience shows that it is a much harder nut to crack.

Dissection into a triangle happens to be simplest, precisely because there is a chord at an angle of 60° to a side. Profiting by this, we can cut the nine-gon into six pieces and rearrange them as in Fig. 16.1, getting a polygon consisting of a triangle

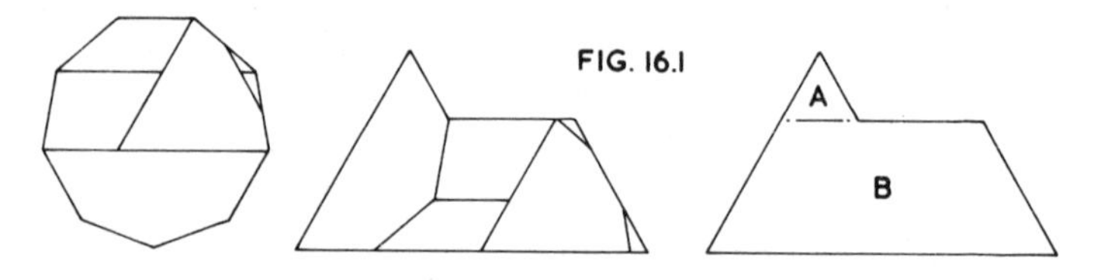

FIG. 16.1

A perched on a trapezoid *B*. Now a *Q*-slide as described with reference to Fig. 1.10 will change the trapezoid into another one with the same angles, whose top side is equal to the side of triangle *A*. Thus, the *Q*-slide changes *A* plus *B* into an equilateral triangle—the triangle we are after. In Fig. 16.2, *CD* is equal to

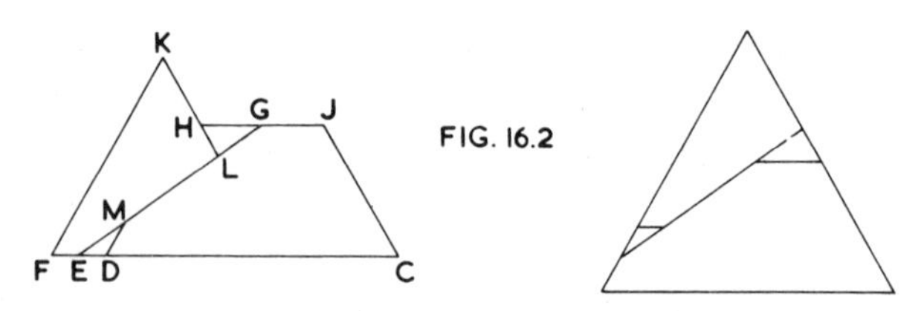

FIG. 16.2

the side of the equivalent triangle, *E* is the midpoint of *DF*, *G* is the midpoint of *HJ*, *HL* is *KH* produced, and *DM* is parallel to *FK*. The complete dissection of nine-gon and triangle, obtained by combining the cuts in Figs. 16.1 and 16.2, is shown in Fig. 16.3.

There is another way of getting the outline of Fig. 16.1, which requires seven pieces instead of six. But it has a degree of freedom that enables us to get an alternative dissection into a triangle with the same number of pieces as in Fig. 16.3. In Fig.

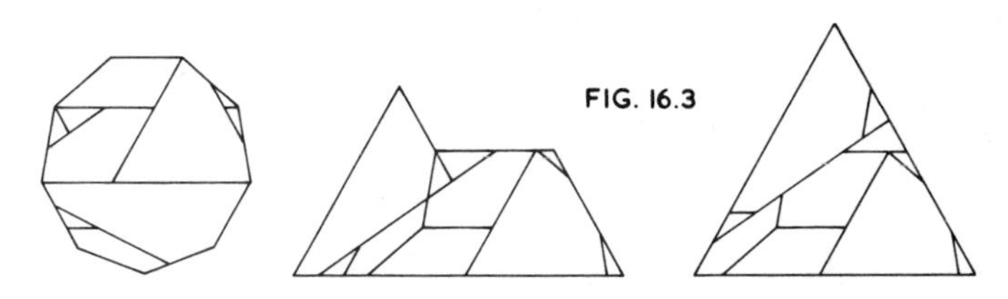
FIG. 16.3

16.4 line *N* can terminate anywhere on *OP,* so the position of line *Q* can be varied appreciably. We shall arrange that *RS* is equal to the side of the small triangle; it will be found that this occurs if *T* is made equal to *U*. If we now put the small triangle on *RS* and add the same cuts as before, we find, as in Fig. 16.5, that the cut *HL* of Fig. 16.2 is *already there.* This offsets the extra piece in the preliminary step.

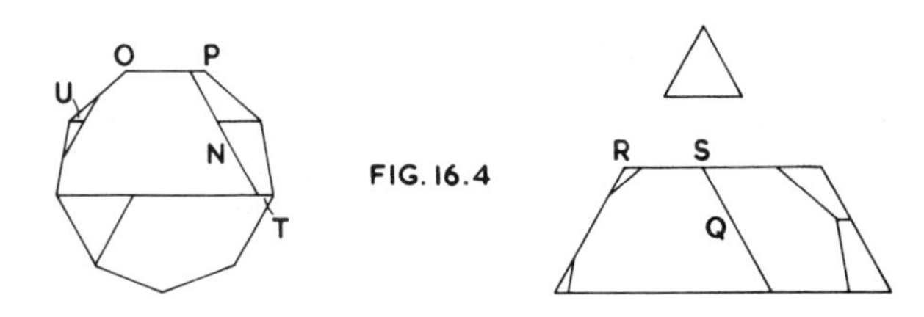

FIG. 16.4

The triangle-trapezoid combination in Fig. 16.1 is a tessellation element. (There is a moral here. Always examine whether a polygon is a tessellation element, for you never know.) But in this case the tessellation does not lead to economical dissections, and the same applies to the tessellation element in Fig. 16.6, even though it has only four pieces.

Better results are obtained by proceeding as in Fig. 16.7, which shows a five-piece rearrangement into a combination of two trapezoids. Here dimension *V* is at our disposal, thanks to which we can make the rearrangement into a tessellation element. Examination of Fig. B9 will show that it is a tessellation because the

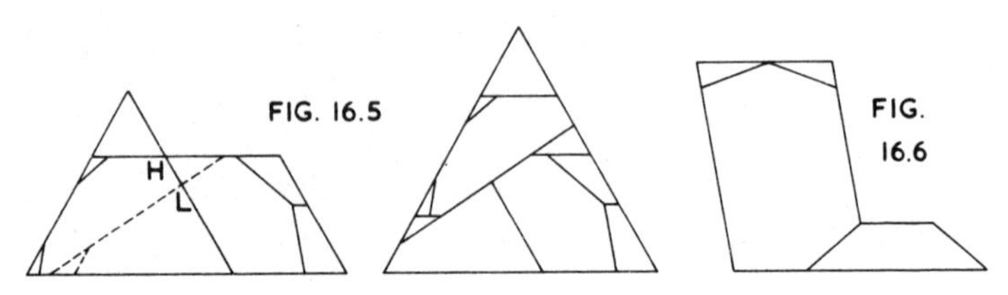

FIG. 16.5

FIG. 16.6

mean lengths WX and YZ of the trapezoids are equal. Dividing an area of 20 square units by the height of the two trapezoids in Fig. 16.7, which can be shown to be ½ tan 20° times the side

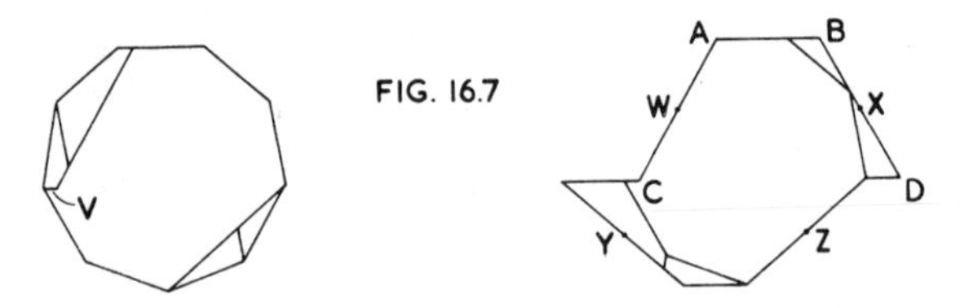

FIG. 16.7

of the nine-gon, we find that its mean width is 3.9212. The mean of AB and CD is found from Appendix C to be

$$-V + s + s \cos 20° + s \sin 20°/\sin 60° = -V + 4.1993,$$

whence

$$V = 4.1993 - 3.9212 = 0.2781.$$

With this value of V we get the tessellation in Fig. B9.

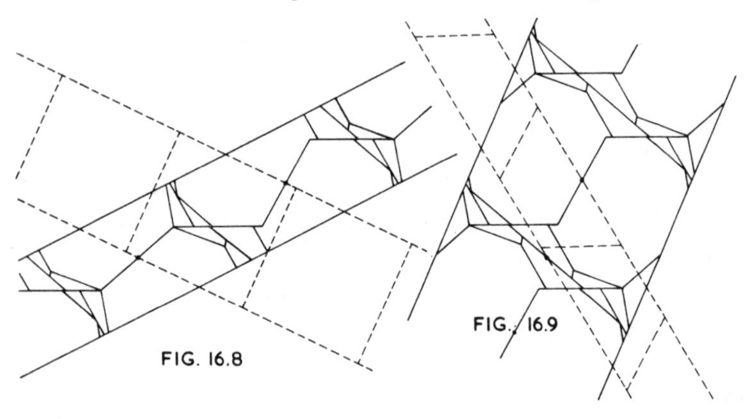
FIG. 16.8

FIG. 16.9

The T-strip we use is obtained from the tessellation by a line through YX and one parallel to it. Figure 16.8 shows how we can use this strip to get a square.

Quite a number of strips can be drawn on the tessellation, but when we come to use them for dissections, we find that the numbers of pieces are depressingly large. The strip most successful with the hexagon is obtained by drawing a line through ZX in Fig. 16.7 and one parallel to it, spaced so as to give a T-strip. The superposition is shown in Fig. 16.9.

A Dissection by Irving Freese. A dissection of nine-gon and triangle in only nine pieces came to my notice after this book was written. It is so simple that I deserve censure for missing it.

The procedure is to assemble the three trapezoids in Fig. 16.10

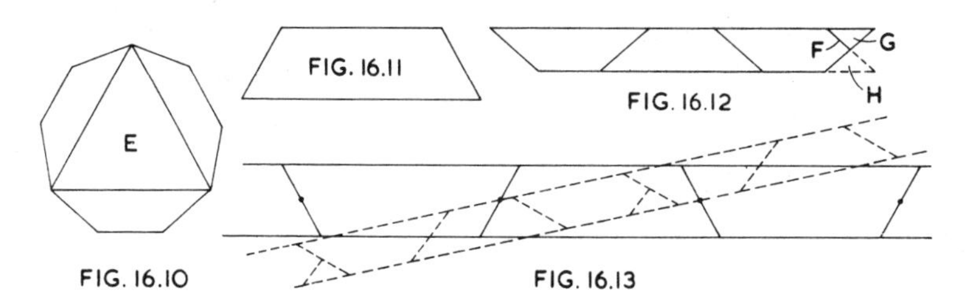

FIG. 16.10 FIG. 16.11 FIG. 16.12 FIG. 16.13

by a PT dissection into the single one shown in Fig. 16.11; the latter combines with triangle E to form the triangle sought. The three trapezoids, joined end to end as in Fig. 16.12, form the element of a T-strip, but the minimal dissection requires a P-strip, whose element is obtained by adding cut F and transferring piece G to H. The parallelogram has an eight-piece dissection into Fig. 16.11, shown in Fig. 16.13.

I tried all possible variants of S- and Q-dissections of the trapezoids in Figs. 16.10 and 16.11, seeking to reduce the number of pieces to seven, but in vain. The conclusion is that if you can beat Freese, you will have found a needle in a haystack.

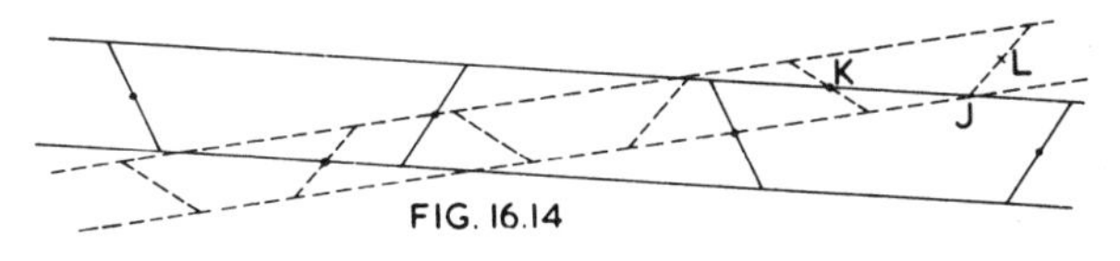

FIG. 16.14

One of the variants tried is nearly successful. Leave Fig. 16.12 as a T-strip element, and superpose it as in Fig. 16.14. This would give an eight-piece dissection of nine-gon and triangle, if only vertex J in the superposed strip were above the upper edge of the other strip. But calculation shows that the angle between strips is 11° 46′, and that ∠*JKL*, instead of being less than this as we would like, is 13° 5′. In other dissections (e.g. Figs. 3.8 and 6.2) the close shave is in our favor, but here it is against us. There is, however, the consolation that the ninth piece is a mere speck, and if we ignore it we have an approximate eight-piece dissection.

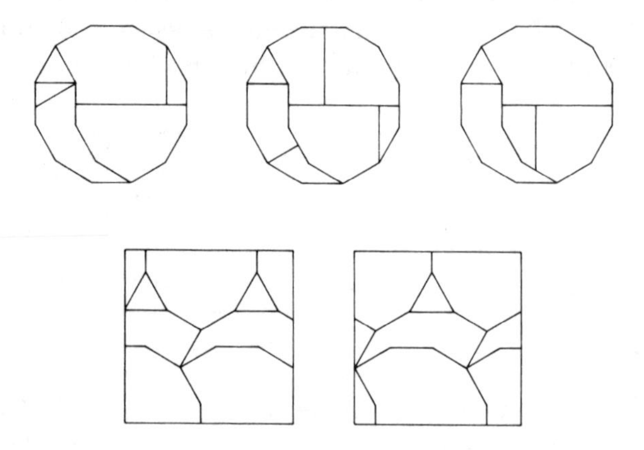

17

THE DECAGON

The decagon is much more satisfying to dissect than the nine-gon. With the latter nothing goes right, but with the decagon we find at the outset that there are several ways of getting useful tessellations. The three made use of are those in Figs. B10, B11, and 17.13.*

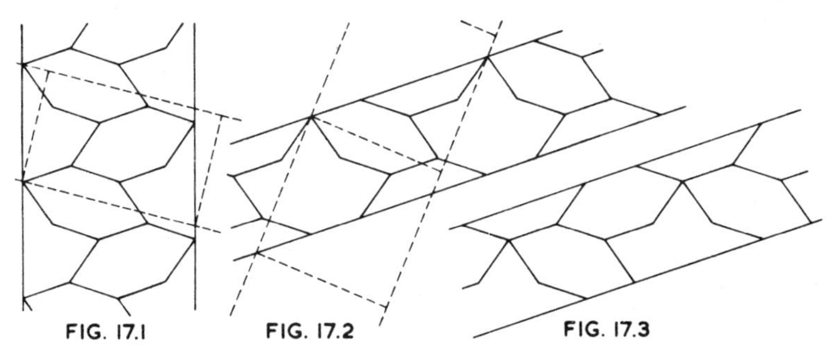
FIG. 17.1 FIG. 17.2 FIG. 17.3

Strips are obtained from the first tessellation by drawing congruent lines on it that are either vertical, or inclined to the horizontal at an angle of 18° either way. (The strips with lines sloping up and with lines sloping down differ only in that one is the mirror image of the other.) The strip obtained with vertical congruent lines can be used, as in Fig. 17.1, to get a dissection into a golden rectangle. One of the strips obtained with sloping lines is used in Fig. 17.2 to get a dissection into a square. Another is shown in Fig. 17.3.

Strips can be obtained from the second tessellation by drawing horizontal lines on it, and only such strips are used here. The lines are drawn through the small hexagons, leaving the large pieces uncut; the reason for this was explained in connection

* See also Appendix G.

with Figs. 3.5 and 3.6. Also, the lines as drawn in Fig. 17.4 pass through the centers of the hexagons. But it will be borne in mind that a dissection may be improved by shifting the lines up or down a little.

It looks as if one gets a different tessellation, if the concavities in the large pieces face right instead of left. But it is just the same tessellation inverted.

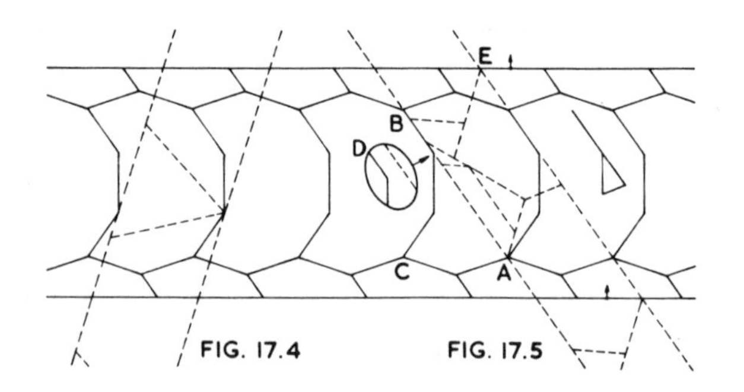

FIG. 17.4 FIG. 17.5

As this strip is rather wide, the strip superposed on it must be rather narrow. Thus, the triangle *T*-strip is far too wide, but the *P*-strip obtained by bisecting it is narrow enough, and in fact gives a minimal dissection of decagon and triangle, as in Fig. 17.4.

The same strip can be used for the more ambitious undertaking of dissecting the decagon into a heptagon. To do this in a small number of pieces is expecting too much; Fig. 17.5 shows the most economical dissection I have found, which requires 14 pieces. (But I don't like the piece exhibiting the Gibbs phenomenon, shown separately on the right.) The heptagon strip is the one in Fig. 15.5, superposed so that vertices in the two strips coincide at *A*. The angle between strips is found to be 54° 13′ and $\angle BAC$ in the decagon is 54°, so each edge of the heptagon strip is just clear of the nearby side *D* of the decagon, as shown in the inset. Vertices almost coincide at *E;* to put beyond doubt that there is no crossing of cuts, the edges of the decagon strip can be raised a little as indicated by the vertical arrows.

The uses of the tessellation in Fig. B10 may not be confined to dissections, but may extend to finer arts. It is fascinating, and its beauty is enhanced by coloring it, preferably using three colors. It has a family resemblance to the tessellation with element as in Fig. 9.9.

The tessellation in Fig. B11 also is a member of a family (but only a small family). Another member is the tessellation in Fig. B8, and the hexagon tessellation could be included as the simplest member.

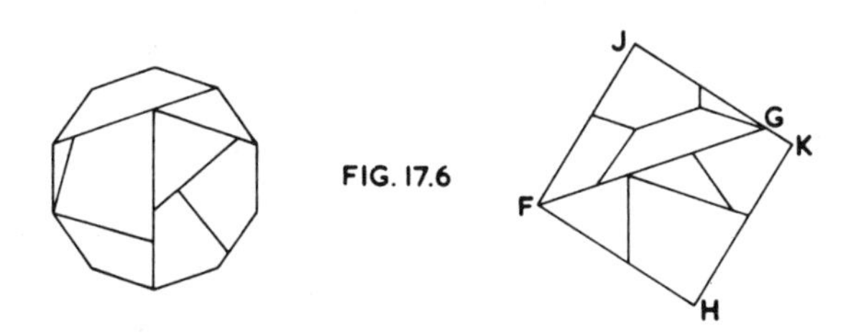

FIG. 17.6

A Dissection by Irving Freese. When the typescript and drawings for this book were almost finished, Dr. C. D. Langford sent me Freese's unpublished decagon-square dissection shown in Fig. 17.6. It will be interesting to see how it fits in with the methods using strips and tessellations.

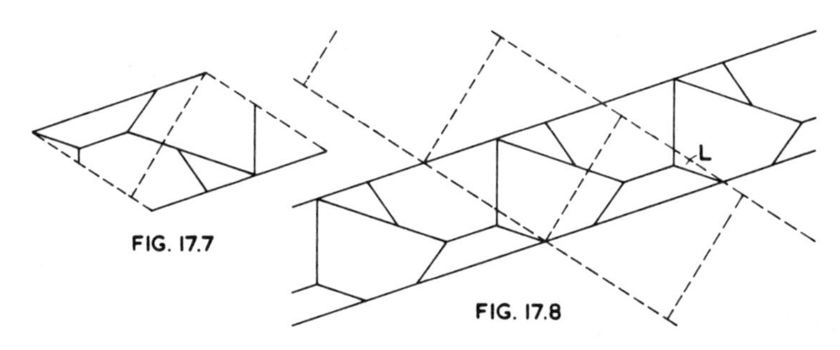

FIG. 17.7

FIG. 17.8

It is characteristic of a square (as well as other simple polygons) that when its dissection is obtained by superposing strips, there is a line such as *FG* extending right across the square. This line is of course one edge of the decagon strip; its other edge passes through *H*, while *FH* is an edge of the superposed strip of squares. Hence, we retrieve the common area by shifting $\triangle FGJ$ to the position in which *FJ* lies along *HK*. The common area is shown in Fig. 17.7, lines belonging to the superposed strip of squares being broken as usual. We need only continue the pattern of full lines on each side of the common area, to retrieve the decagon strip and the *PP* superposition in Fig. 17.8. It is now clear that piece *L* can be made larger, more on a par with the other pieces, by displacing the strip of squares as in Fig. 17.9. In respect of the sizes of the pieces, this dissection is distinctly better than Fig. 17.2.

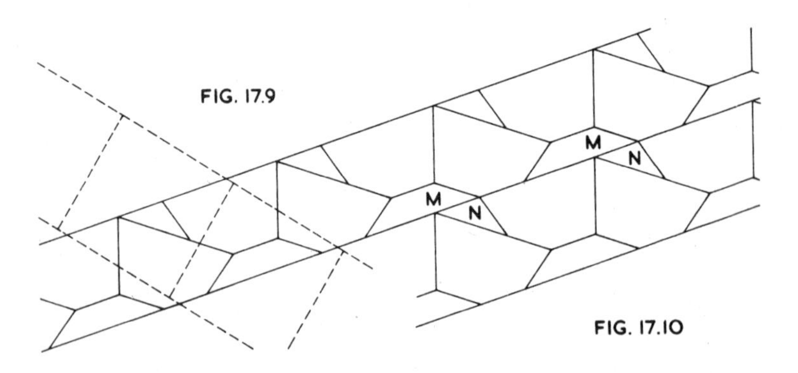

FIG. 17.9

FIG. 17.10

The question now arises, is this decagon strip derivable from a tessellation? If so, the tessellation consists of strips laid side by side with the line pattern of each in suitable relation to those of its neighbors. To find the relative position, we slide one strip along its neighbor until the position of some piece on the upper edge of the lower strip, relative to that of some piece on the lower edge of the upper strip, is the same as in Fig. 17.6. This is found to be the case with pieces *M* and *N* in Fig. 17.10. Therefore, the line between *M* and *N* is not part of the tessellation, but part of one of the congruent lines bounding a strip. Deleting it, we get the cuts in Fig. 17.11 to make the decagon a tessellation element. The element itself is shown in Fig. 17.12, and the tessellation in Fig. 17.13.

Using the strip deduced from Freese's dissection, I found three dissections that are more economical than any using strips derived from Figs. B10 and 11. Those of hexagram and Greek cross appear in Figs. 17.14 and 17.15. The third is the dissection of a Latin cross in Fig. D36.

Other strips are obtained from Fig. 17.13 by drawing lines through *OP, QR* and *OQ, PR*. Several of the dissections they yield

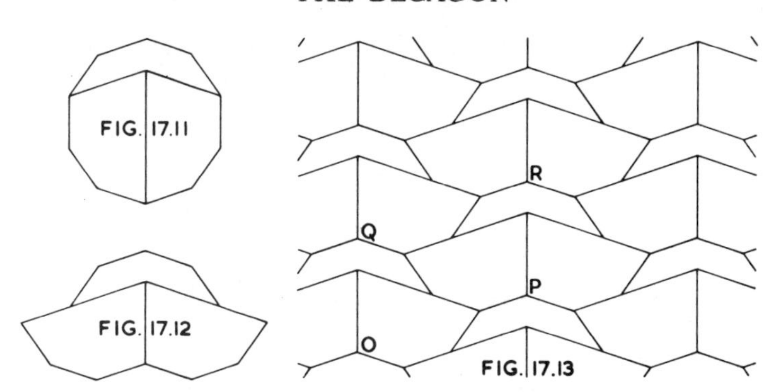

FIG. 17.11
FIG. 17.12
FIG. 17.13

are as good as those from other strips, but none appears to be better.

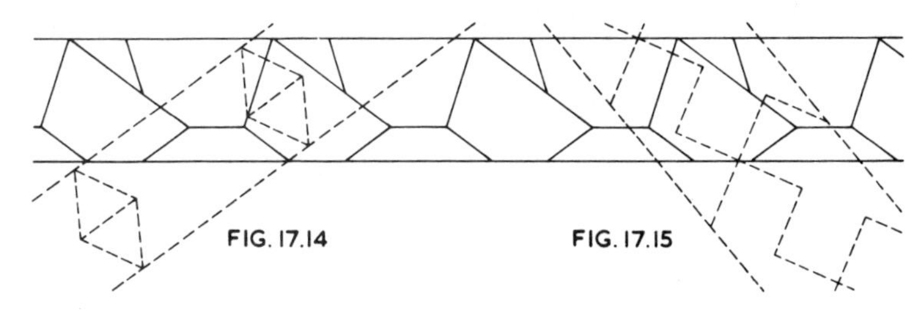
FIG. 17.14
FIG. 17.15

This episode of Freese's decagon-square dissection demonstrates that two heads can be better than one.

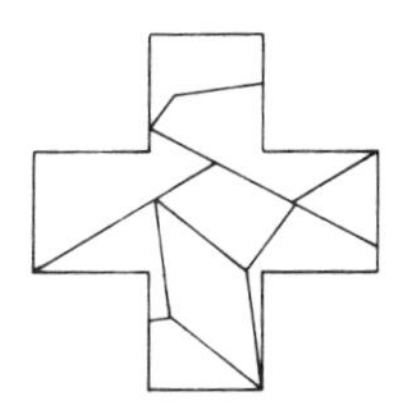
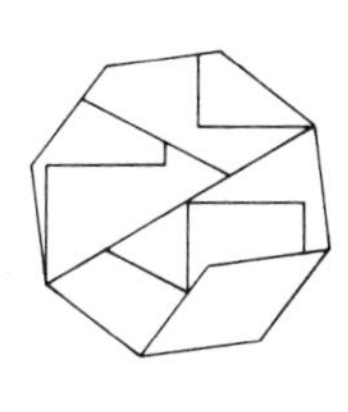

18

RECTILINEAR LETTERS

Seeking fresh fields to conquer, we turn to the rectilinear letters. Curvilinear letters are excluded because they can be dissected only if they are excessively distorted, so as to get edges including convex and concave curves that fit together. We confine ourselves almost entirely to dissections into a square, and to the more tractable letters.

FIG. 18.1

There seems to be a tacit agreement that letters for dissection shall have a height that is five times the thickness of the strokes. Accordingly, the letters attacked will be dimensioned as in Fig. 18.1. Although they look true to type, most of them have been modified slightly to simplify dissection. For instance, the sloping strokes have the same width as other strokes only if we measure the width horizontally. This uniform width will be our unit of length throughout.

It is convenient to use graph paper in dissecting letters, and not to make their area a constant 20 sq. cm. or other value. Instead draw them on any scale that suits the paper, and if a superposition is to be tried, draw a separate strip of squares for the letter (using the graph paper as a guide, of course).

Some of the simplest letters are easily squared by means of a

P-slide. The simplest of all was done in Fig. 1.2. Others are dealt with in Figs. 18.2-18.4.

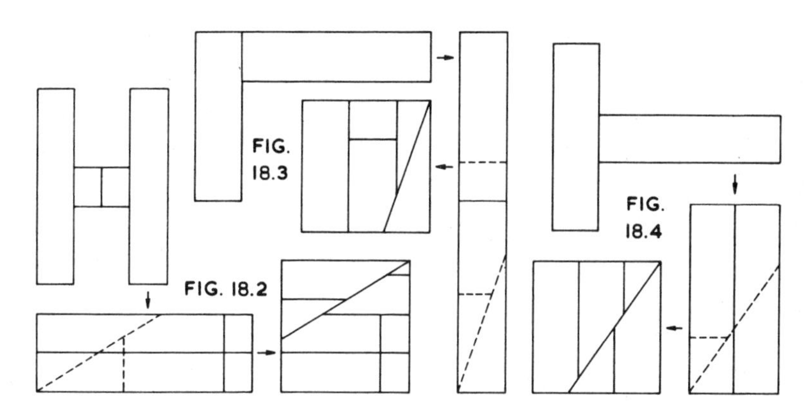

An N also can be squared by a *P*-slide, but the preliminary rearrangement is not so immediate. The area of the N is $13\frac{1}{3}$ square units, i.e., $5 \times 2\frac{2}{3}$, and as large parts of the N are 5 units long, there is a good chance of getting an economical strip element 5 units long and $2\frac{2}{3}$ units wide. In fact trial soon yields the element in Fig. 18.5, which is easily made into a rectangle.

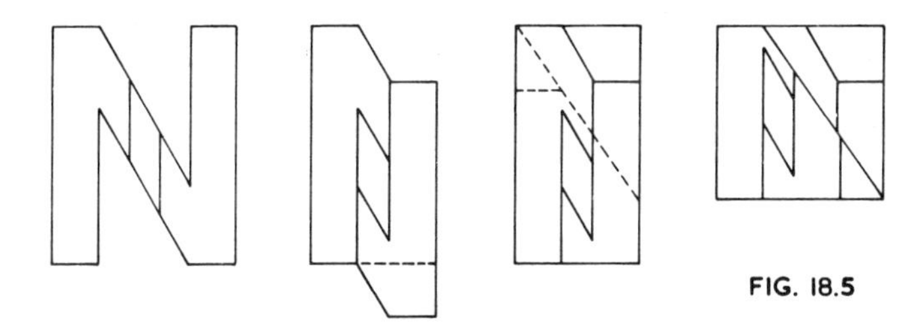

FIG. 18.5

The *P*-slide cuts applied to the rectangle cross only one existing cut, so we have quite an economical dissection of N and square.

The manner of attack depends to some extent on the value of the area, as has just been seen. The area of some letters is a perfect square, e.g., M, X, Z have areas of 16, 9, 9 square units respectively. When this is the case, it may be easy enough to proceed direct from letter to square by trial. Thus, a 4×4 square *ABCD* can be drawn on the M where shown in Fig. 18.6, and one then tries filling the empty spaces inside *ABCD* with

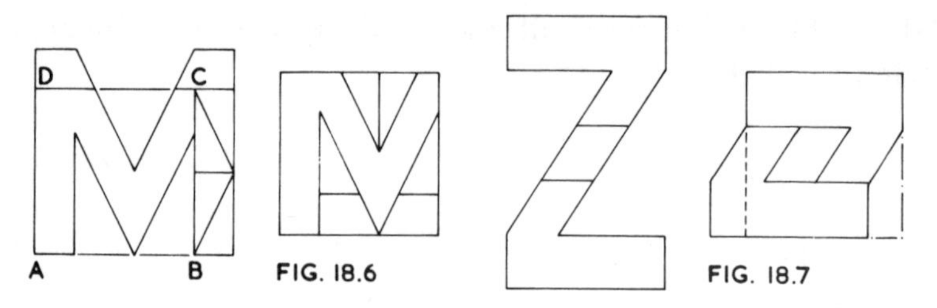

FIG. 18.6

FIG. 18.7

pieces cut from the parts of the M outside. Similarly with Z, as in Fig. 18.7, but with this letter we take a hint from Fig. 18.5.

Since X has only sloping strokes, one does not try for a square direct, but for a parallelogram 3 units long and 3 units high. This is done in Fig. 18.8 by turning over pieces *E* and *F* so that they join on to piece *G* to form a long, narrow parallelogram, chopping the latter into three equal pieces, and stacking them. Two vertical cuts in the stacked parallelogram give a square. Pieces turned over are marked with the encircled dot and cross familiar to students of electricity. If you scorn the ruse of turning over, a further cut, and, therefore, a further piece, is required, in order that *E* plus *F* may join on to *G*.

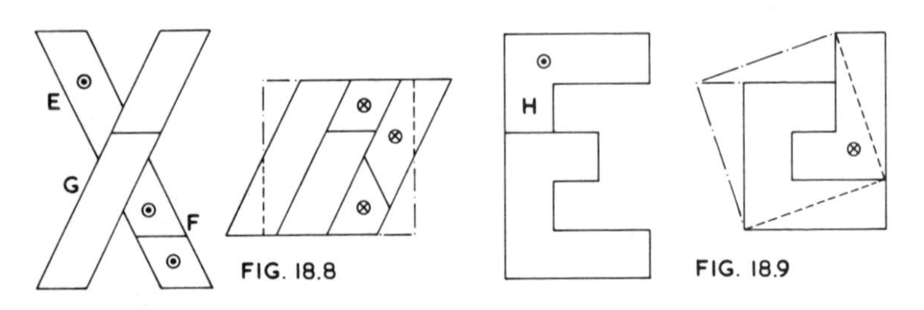

FIG. 18.8

FIG. 18.9

The area of a letter may not be a perfect square, but instead the sum of two squares. Thus, the area of E is 10, i.e., $9 + 1$, to which arithmetic fact corresponds Dudeney's geometric rearrangement in Fig. 18.9. The cuts giving the square are copied from Fig. 9.17. Turning over can be avoided, at the expense of an extra piece, by making the small square *H* a separate piece.

Similarly the area of F is 8, which bright sparks will recognize as twice 4. This indicates that you should try superposing a tessellation of squares of side $2\sqrt{2}$, keeping your eyes open. Several five-piece solutions can be found. The one in Fig. 18.10

was chosen because the shapes of the pieces are least suggestive of the letter dissected.

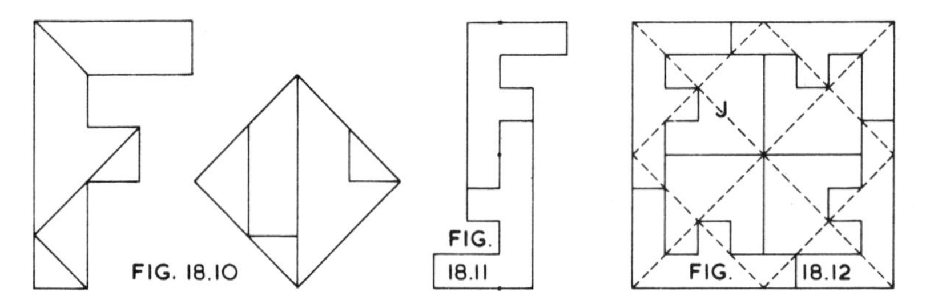

FIG. 18.10 FIG. 18.11 FIG. 18.12

This F happens to be the element of a square tessellation, half of whose elements are inverted as in Fig. 18.11. Superposing a tessellation of squares with their vertices on the centers of symmetry marked by dots, we get another five-piece dissection, alternative to Fig. 18.10. The latter also is related to a tessellation, only it happens to be one of the cases in which it is easier to find the tessellation from the dissection than vice versa. It can be deduced from Fig. 18.10 that the tessellation has two elements, namely the F and piece *J* in Fig. 18.12. The tessellation is of the kinds discussed in Chapters 11 and 12. Superposing a tessellation of squares gives the five-piece dissection of the F and a four-piece one of element *J*.

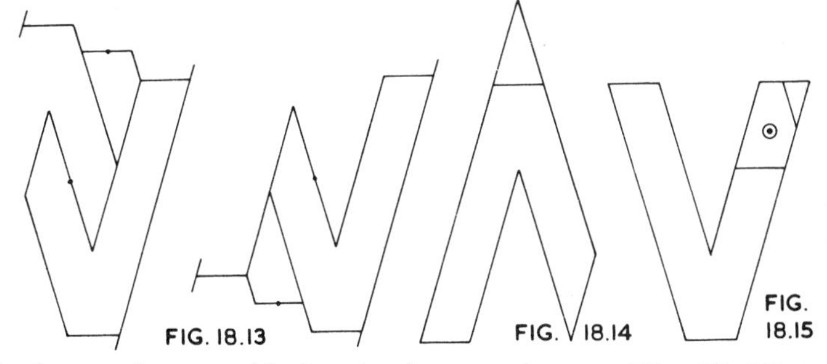

FIG. 18.13 FIG. 18.14 FIG. 18.15

So far we have tackled only the easy letters. The V, W, and Y that remain will obviously not easily succumb to a *P*-slide, and their areas, which are respectively 55/6, 65/4, and 15/2, are not squares. Certainly 65/4 is a sum of two squares, and in two ways to boot, but this is a blind alley. So we try general methods based on tessellations and strips.

Two V's, one inverted, can be joined to form a kind of N or Z. Again taking the hint from Fig. 18.5, we find from two joined V's the *T*-strip elements in Fig. 18.13. A search for variants yields the tessellation element in Fig. 18.14. Any of these three will be found to give an eight-piece dissection, but we go one better with the further variant formed by cutting as in Fig. 18.15. Rearrangement as in Fig. 18.16 gives a *T*-strip from which we find a seven-piece *TT*2 dissection. To avoid turning over, we can cut the piece turned over into a hexagon and triangle.

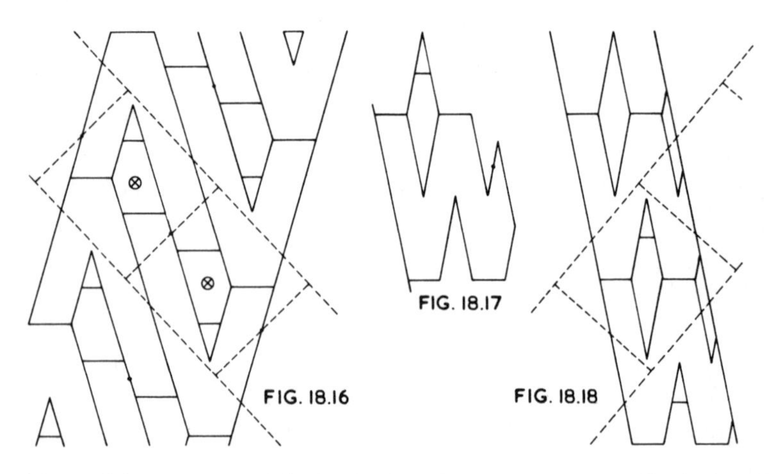
FIG. 18.16 FIG. 18.17 FIG. 18.18

After this success with V, it is relatively easy to find that W has a *T*-strip element Fig. 18.17. But this time a better dissection is found by bisecting the *T*-strip to get a *P*-strip.

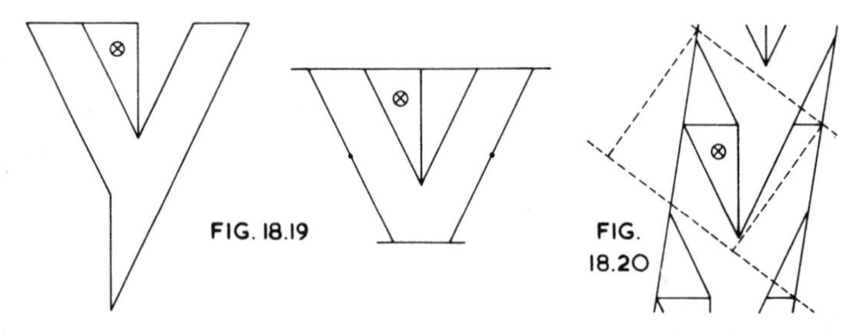
FIG. 18.19 FIG. 18.20

Y is made a tessellation element as on the left of Fig. 18.19. It can be seen immediately that there is a *T*-strip element as shown, and an eight-piece dissection can be found from it. But on second thoughts we draw the tessellation, find from it the *P*-strip as in Fig. 18.20, and get a seven-piece dissection.

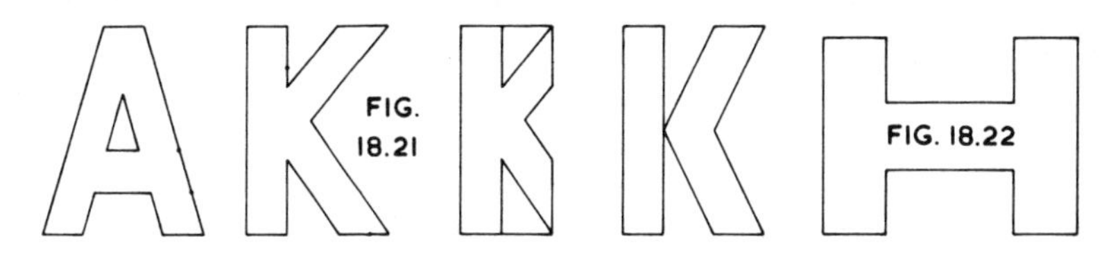
FIG. 18.21
FIG. 18.22

The only rectilinear letters not dissected are the problem ones A and K. It appears that letters reasonably true to type, as on the left of Fig. 18.21, are best dissected piecemeal. But this requires too many pieces, and the promising alternative of first making the K into a notched rectangle promptly comes to nothing. On the other hand easily dissected shapes like the K on the right are too distorted. So I leave the dissection of A and K to someone else.

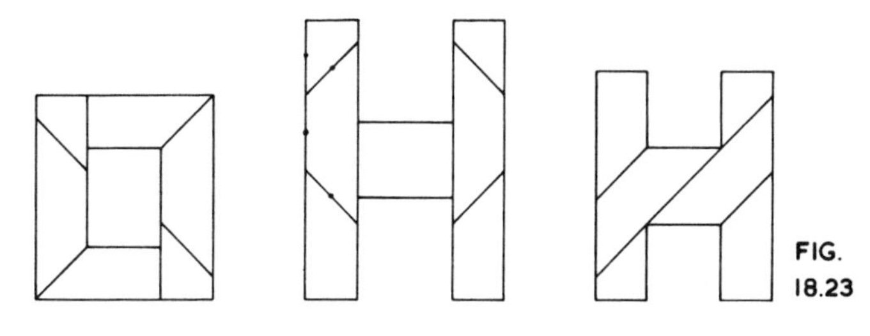
FIG. 18.23

There are two dissections of an H that are worth mentioning, even though the H's are rather distorted. The first H is made up of seven pieces as in the middle of Fig. 18.23. Each line marked with a dot at its midpoint is $\sqrt{2}$ times the width of the vertical strokes. The dimensions of the cross-piece are immaterial, for the problem is to remove it and form an H with the remaining six pieces.

The second H is the squat one in Fig. 18.22, which is readily dissected because it is the element of a square tessellation. Joseph S. Madachy's dissection into a square appears in Fig. 18.24, another in Fig. 18.25. The dissections into one Greek cross and into two are the result of superposing tessellations of Greek crosses.

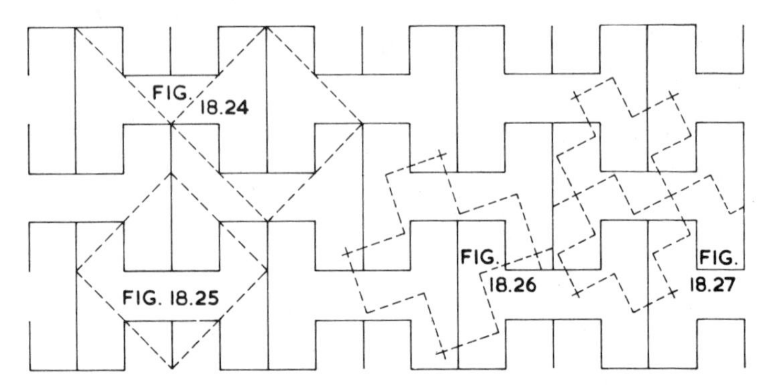

One can wander further afield and explore foreign alphabets. Dissections of gamma, lambda, and the Russian soft *i* in Fig. 18.28 are obvious from those of L, V, and N, and the dissection of xi is similar to that of H, with seven pieces instead of eight.

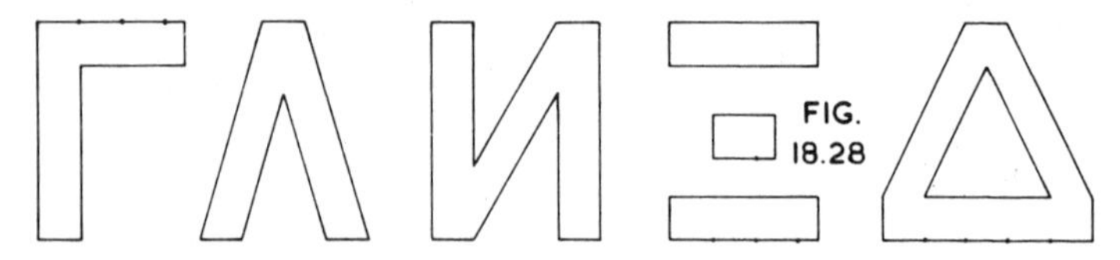

But delta is a more fascinating problem. Shaped as in Fig. 18.28, its area is 25/2 and the length of its base is 5. These dimensions should make possible a *T*-strip element of mean width 5/2, with the bases of deltas forming its edges as on the left of Fig. 18.29. But the top of the delta would be in the way, so we cut it off and put it aside for the time being. We want the left-hand side of the element to be symmetric about the dot; this is most easily effected by placing the edge of the base of an inverted element at *K*. Adding an inverted element in this position, we find overlaps in the spaces marked *L*, and spaces still empty where marked *M*. The obvious remedy yields the strip in Fig. 18.30, from which we get an eight-piece dissection.

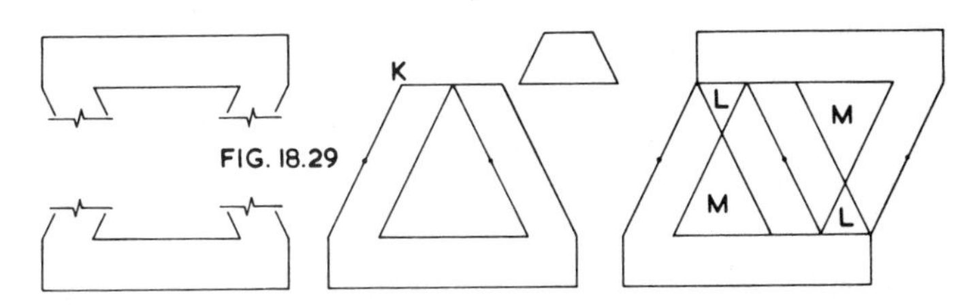

FIG. 18.29

Such a small number of pieces was a pleasant surprise. Another one was the result of superposing a strip of triangles on the same strip. This *TT*2 dissection has only seven pieces.

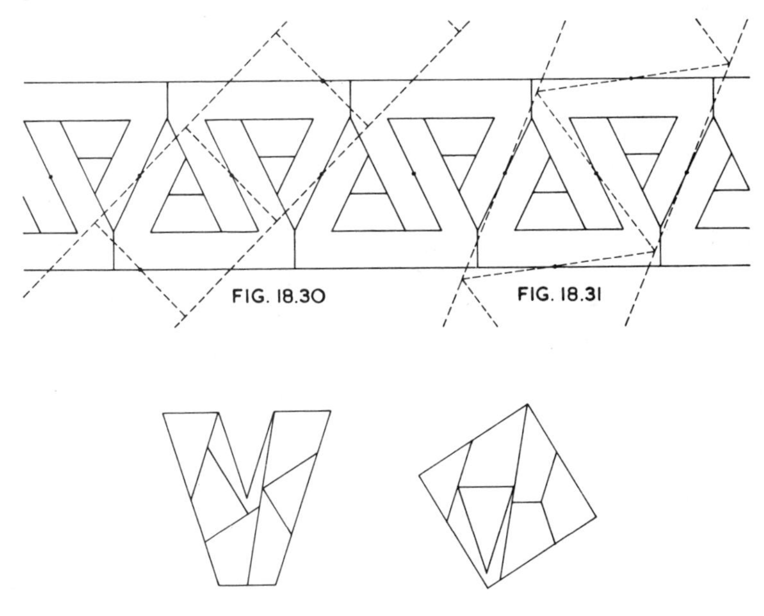
FIG. 18.30 FIG. 18.31

19

STARS

A piecemeal dissection of pentagram and square is given in Fig. 6.5. But the scope of piecemeal dissections is limited, so we try the procedure via tessellations and strips.

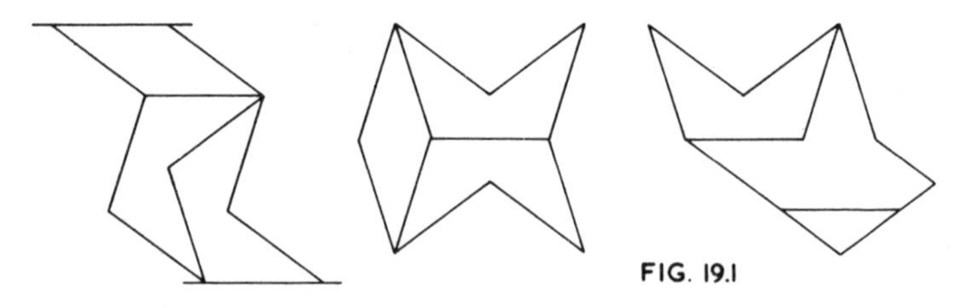

FIG. 19.1

Simple ways of getting tessellation elements from a pentagram are shown in Fig. 19.1. The first is already a *P*- and *T*-strip element, and as it is rather wide, we invert alternate elements and get a strip suitable for *TT*2. (See the discussion of Fig. 5.3.) With this strip we get a dissection into a hexagon as in Fig. 19.2.

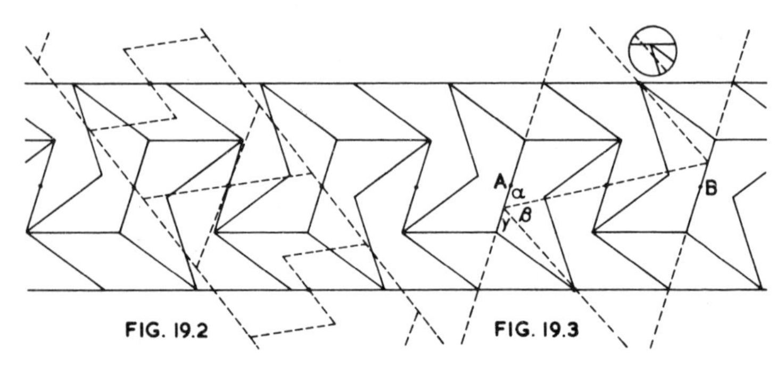

FIG. 19.2

FIG. 19.3

A dissection into a triangle can also be obtained. Figure 19.3 shows the result of superposing, drawn as if the angle between strips were 72° and not 73° 48′. Since the exact angle is greater than 72°, the sides of the triangle strip should be slightly rotated anticlockwise about A and B, narrowing it and introducing a minute piece as in the inset. But if we leave said sides as they are, we can draw triangles between them in which $\alpha = 60° \; 54'$ and $\beta = \gamma = 59° \; 33'$. Thus, we have an approximate seven-piece dissection of pentagram and equilateral triangle. The exact dissection requires two more pieces, one being the minute piece shown in the inset, the other being the thorn in the side of the pentagram produced by the rotation. The size of the first extra piece and the shape of the second constitute blemishes; a better nine-piece dissection would be desirable.

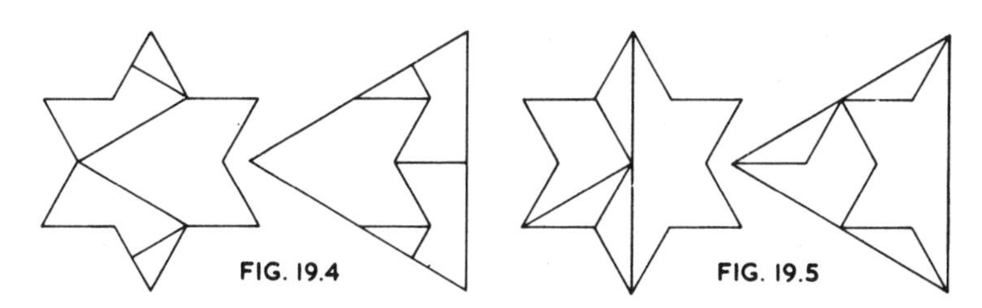

FIG. 19.4 FIG. 19.5

The hexagram is as compliant as the pentagram is obstinate, and has already featured in several dissections. Two more are given in Figs. 19.4 and 19.5. They are classifiable as S- or T-dissections, but the metric relations between hexagram and triangle are so simple that strips and tessellations need not be called in aid.

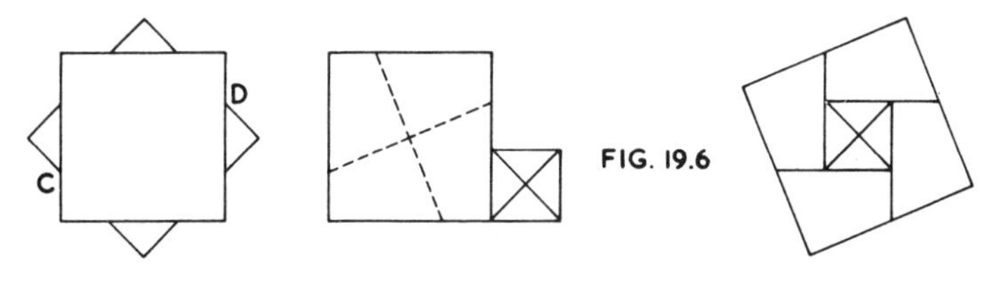

FIG. 19.6

Of the other polygrams * we consider in this chapter only the octagram {8/2}, that is, the one formed by joining every second vertex of an octagon. It is easily changed into two squares and thereupon into one (Fig. 9.17 again). Other dissections into a square can be found by trial, based on the fact, evident from

* One says "polygon" and "n-gon", so why not "polygram" and "n-gram"? But remember that the latter has *nothing* to do with dianetics! (Martin Gardner, *Fads and Fallacies,* Chapter 22.)

Fig. 19.6, that the side of the equivalent square is equal to CD. But none is so symmetric as this one.

The S- and T-methods can always be applied to stars, but I confess to a lack of interest in most of these dissections. More interesting are those of a star and another polygon whose dimensions are simply related, by virtue of which the number of pieces is fairly small. These are taken up in Chapter 20.

A problem that naturally suggests itself is to dissect an n-gram into an n-gon, because their angles are simply related. There is, however, in general no simple relation between their linear dimensions, and this precludes a simple general dissection. Failing this, we can seek a simple general dissection of an n-point star that is not an n-gram. It is of course easier to start with the n-gon and see what star we can dissect it into.

There is, in fact, a simple way of doing this, which will be explained with reference to Fig. 19.7. Since the central triangular piece in this figure has half the side of the triangle to be dissected, as will be shown presently, the cuts are most conveniently found by drawing the central piece first. Draw a circle whose center is the center of the small triangle and whose radius is twice its circumradius. Then produce the sides of the triangle to meet the circle at E, F, G, and join these points. This construction applies to any n-gon.

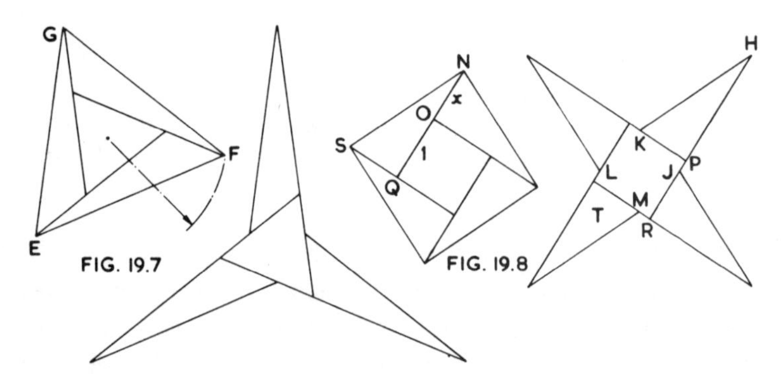

FIG. 19.7

FIG. 19.8

The proof of the construction will be illustrated by the case $n = 4$, but is quite general. Let the n-gon to be dissected have side 2, and let the small n-gon have side 1. Then, referring to

Fig. 19.8, it suffices to prove that $HJ = HK$. Each point of the star will then be an isosceles triangle erected on a side of the regular n-gon formed by joining $JKLM$, so the star will be symmetric as required.

Let $NO = x$. Then

$$HP = NQ = x + 1,$$
$$HJ = HP + PR - JR = x + 1 + 1 - x = 2 = NS = HK,$$

and it follows that the star is symmetric. If you want to do the dissection the hard way, starting with the n-gon to be dissected, you can show that

$$x = \tfrac{1}{2}[\sqrt{\{3 \csc^2 (\pi/n) + 1\}} - 1],$$

by applying the cosine formula to $\triangle NQS$.

These n-point stars can be made in the form of hinged models. Thus, each $\triangle NQS$ in Fig. 19.8 can be hinged to the central square at Q, and rotated anticlockwise about Q into position T.

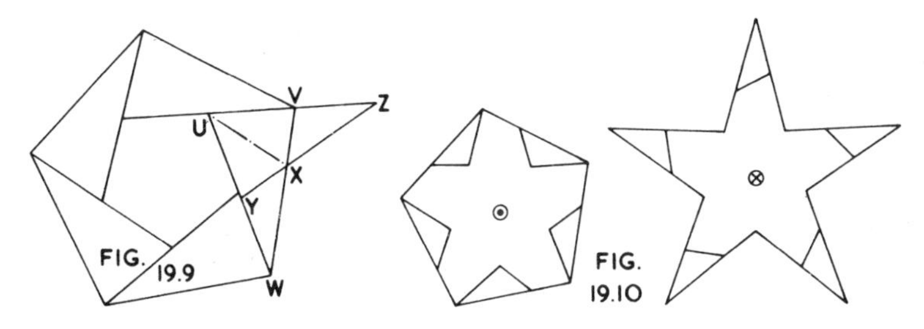

A modified dissection can be derived from this one. Having cut a pentagon into a small one plus five triangles as in Fig. 19.9, do not rotate $\triangle UVW$ about U as before, but about axis UX into position UYZ. Then triangles WXY and ZXV are mirror images, and the same change in outline is effected by rotating only $\triangle WXY$ about UX. So cut out only the smaller triangles like WXY, rotate them as above, and again you get a five-point star. The dissection is shown in Fig. 19.10, with the single large piece turned over instead of the five small ones. If you try to get a dissection like this but with no turning over, you will find that the facts of geometry won't allow it.

The same idea was tried in order to dissect an n-gon into a

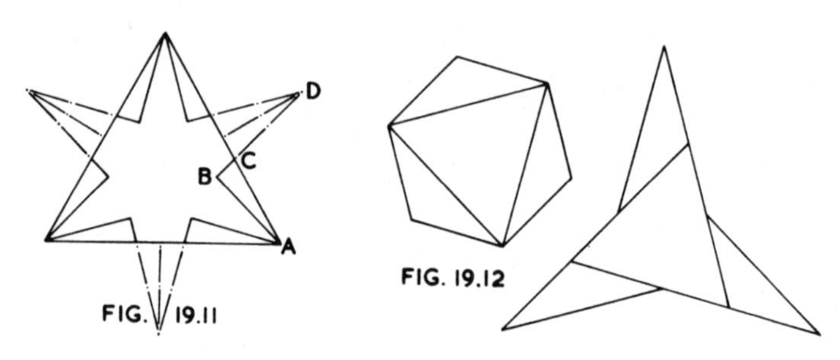

FIG. 19.11

FIG. 19.12

$2n$-point star. But as is seen from Fig. 19.11, the points cannot all have the same shape, for

$$AB < AC = CD < BD.$$

The converse problem of dissecting a $2n$-gon into an n-point star, using methods as described here, succeeds only with the hexagon. We do much better than this in the next chapter, which includes the dissection of any n-gon into a $2n$-gram.

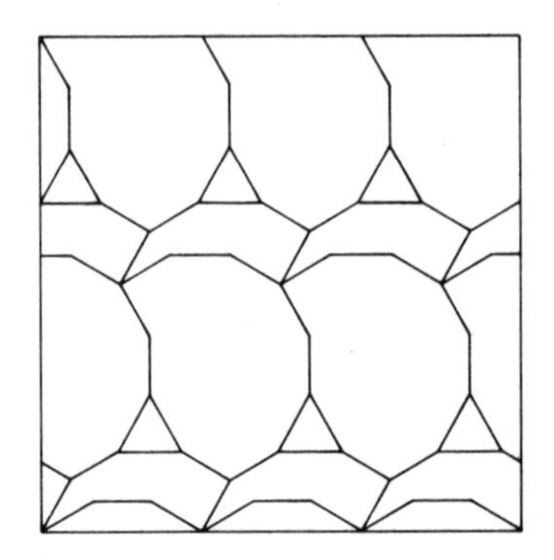

20

MORE STARS

Looking at Figs. 9.7-9.10, 19.5, and 19.6, we see that they have a common characteristic. This is that the end points of every cut in the more complex of the two polygons are determined by intersections of chords. In an n-gon or n-gram the angle between any two chords is π/n or a multiple thereof if n is even, and if n is odd, it can be the complement of such an angle. Hence, any side of any piece in the dissection is equal to the side s_1 of the polygon multiplied by a rational function of sines and cosines of multiples of π/n. (Such a rational trigonometric function will henceforth be called an RTF.) We confine ourselves to pairs of polygons in which the angles π/n are equal, or one is a small integral multiple of the other. In such cases, if the dissection has the abovementioned characteristic, the side s_2 of the polygon obtained by rearrangement is also s_1 times an RTF. The dissection of two polygons so related is unusually economical, as we have seen. This prompts us to make a systematic search for such pairs of polygons.

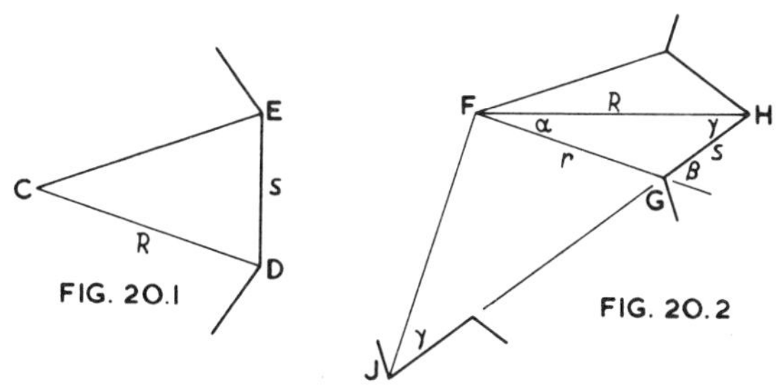

FIG. 20.1

FIG. 20.2

The area A of an n-gon is n times that of the triangle obtained by joining its center C in Fig. 20.1 to vertices D and E. We easily find

$$A = \tfrac{1}{4} ns^2 \cot(\pi/n),\ A = \tfrac{1}{2} nR^2 \sin(2\pi/n),$$

whence

$$s^2 = (4A/n) \tan(\pi/n),\ R^2 = (2A/n) \csc(2\pi/n). \qquad (1)$$

The area of an $\{n/m\}$ is $2n$ times that of the triangle obtained by joining its center F in Fig. 20.2 to vertices G and H. It can be expressed in the three ways

$$Rr \sin\alpha = rs \sin\beta = Rs \sin\gamma = A/n. \qquad (2)$$

We must find α, β, γ. From Fig. 20.2

$$\alpha = \pi/n,$$

and since $\angle HFJ = 2\pi m/n$,

$$\gamma = \tfrac{1}{2}\pi - \pi m/n,\ \beta = \gamma + \alpha = \tfrac{1}{2}\pi - \pi(m-1)/n.$$

Substituting in (2) gives

$$Rr \sin(\pi/n) = rs \cos\{\pi(m-1)/n\} = Rs \cos(\pi m/n) = A/n,$$

and dividing the product of all three expressions by the square of each in turn leads to

$$s^2 = \frac{A \sin(\pi/n)}{n \cos\{\pi(m-1)/n\} \cos(\pi m/n)},$$

$$R^2 = \frac{A \cos\{\pi(m-1)/n\}}{n \cos(\pi m/n) \sin(\pi/n)}, \qquad r^2 = \frac{A \cos(\pi m/n)}{n \sin(\pi/n) \cos\{\pi(m-1)/n\}} \qquad (3)$$

It is just as easy to find formulas for all three dimensions as for s alone.

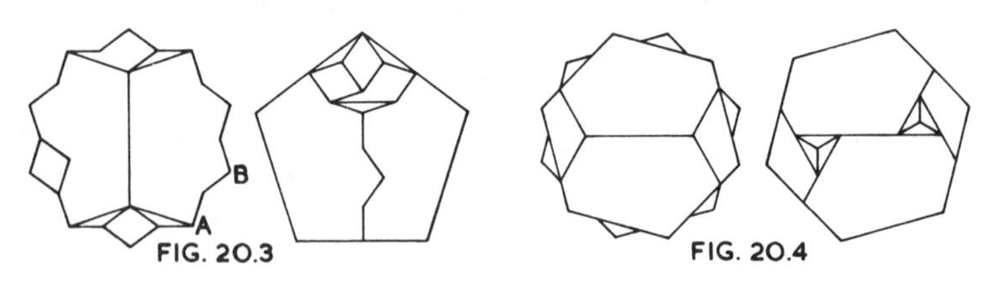

FIG. 20.3 FIG. 20.4

To get back to dissections. If the sides of the polygons, as given by (1) and (3), have as ratio an RTF, then we can expect an economical dissection. Thus, for dodecagon and square we have

$$s\{12\}/s\{4\} = \sqrt{(\tfrac{1}{3} \tan 15^\circ)} = \sqrt{\{\tfrac{1}{3}(2 - \sqrt{3})\}} = (\sqrt{3} - 1)/\sqrt{6}.$$

This is an RTF of 15°, so we might have anticipated Figs. 9.7-9.10. Here the rational relation is more easily discerned from

$$s\{4\}/R\{12\} = \sqrt{(6 \sin 30°)} = \sqrt{3} = 2 \cos 30°. \qquad (4)$$

To get back to stars. The search for dissectable pairs of polygons was conducted by finding quotients of their s's in terms of radicals with $n = 3$-6, 8, 10, 12, and seeing which of them were RTF's. In some cases the trigonometric form of the quotient suggested a general relation of which it is a particular case; in others the dissection, when found, suggested a general relation: and in others none could be found. It is immaterial whether we find a quotient of s's, R's, or r's, or a mixture such as the s/R in (4), since each of the three is an RTF of either of the others.

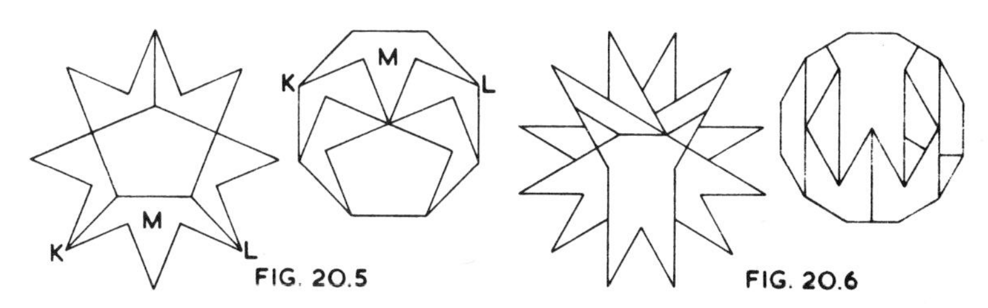

FIG. 20.5 FIG. 20.6

As an illustration of the procedure, we have

$$s\{3\}/s\{6/2\} = 2\sqrt{3},$$

which, as would be expected from Fig. 19.5, is an RTF of 30°. The trigonometric form is

$$\frac{s^2\{3\}}{s^2\{6/2\}} = \frac{4A \sin 60°}{3 \cos 60°} \cdot \frac{6 \cos 60° \cos 30°}{A \sin 30°} = \frac{8 \sin 60° \cos 30°}{\sin 30°}$$

$$= \frac{16 \sin 30° \cos^2 30°}{\sin 30°} = 16 \cos^2 30°,$$

where cos 60° cancels in the second member and sin 30° in the fourth, leaving a perfect square. It is not hard to deduce the general relation

$$\frac{s^2\{n\}}{s^2\{2n/2\}} = \frac{4A \sin(\pi/n)}{n \cos(\pi/n)} \cdot \frac{2n \cos(\pi/n) \cos(\pi/2n)}{A \sin(\pi/2n)}$$

$$= \frac{8 \sin(\pi/n) \cos(\pi/2n)}{\sin(\pi/2n)} = \frac{16 \sin(\pi/2n) \cos^2(\pi/2n)}{\sin(\pi/2n)}$$

$$= 16 \cos^2(\pi/2n),$$

where $\cos(\pi/n)$ cancels in the second member and $\sin(\pi/2n)$ in the fourth. Hence,

$$s\{n\}/s\{2n/2\} = 4 \cos(\pi/2n), \tag{5}$$

that is, the side of any n-gon is twice the distance AB (Fig. 20.3) between adjacent points in the $\{2n/2\}$ of equal area. Dissections based on (5) appear in Figs. 19.5, 19.6, 20.3, and 20.4. In the last of these we use the tessellation in Fig. 9.20, just as in Fig. 19.6 we used Fig. 9.17.

The existence of a relation such as (4) or (5) implies the existence of a dissection, every cut in which is of length an RTF. Such cuts need not be determined in every case by intersections of chords, though most of them will be. Figure 19.4 illustrates this, for two of the cuts therein cannot possibly be determined in that manner, but their lengths are obviously RTF's. On the other hand, it may transpire that a more economical dissection has some cuts whose lengths are not RTF's.

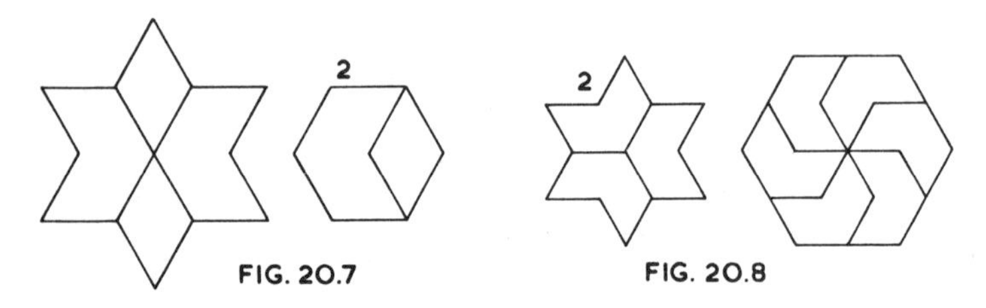

FIG. 20.7

FIG. 20.8

A general relation which, we shall see, was suggested by a dissection is

$$R\{2n\}/s\{2n/n - 1\} = \sqrt{2}. \tag{6}$$

If n is even, $\sqrt{2}$ is an RTF of $\pi/2n$. So we seek and find dissections of $\{8/3\}$ into octagon and $\{12/5\}$ into dodecagon, as in Figs. 20.5 and 20.6. If n can be odd, $\sqrt{2}$ is not always an RTF as required.

But we can still expect a dissection, namely of one polygon into two, for the right-hand side of (6) then becomes 2 or 1. Figures 20.7 and 20.8 are elementary examples.

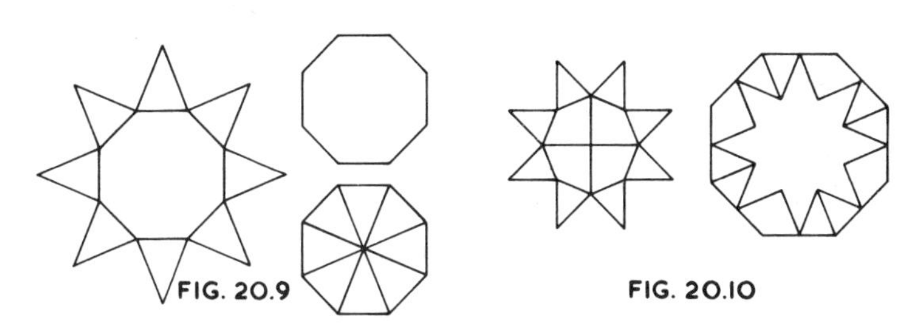

FIG. 20.9 FIG. 20.10

The assembly, predicted by (6), of two $2n$-gons into a $\{2n/n - 1\}$ as in Fig. 20.9 looks trivial. That we get a star is trivial, but that we get a polygram is just a little less so. Besides, it was Fig. 20.9 that led to the general relation (6), and (6) led to Figs. 20.5 and 20.6. There are dissections like Fig. 20.9 for all values of n in (6). They appear to be minimal except when $n = 3$.

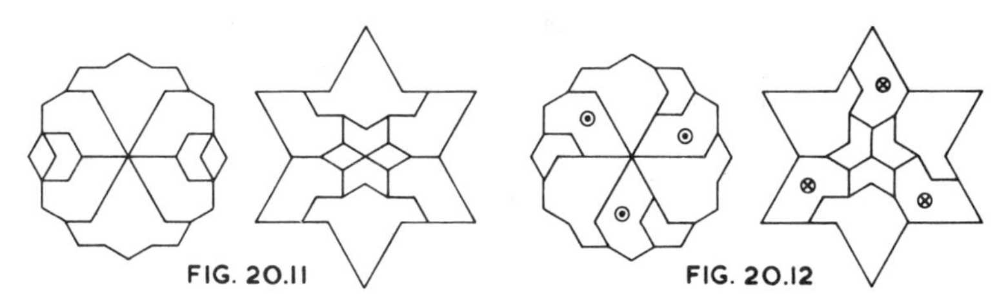

FIG. 20.11 FIG. 20.12

Replace n in (5) by $2n$ and you get

$$s\{2n\}/s\{4n/2\} = 4 \cos (\pi/4n). \tag{7}$$

Dividing this by (6) gives an RTF that can be expressed as

$$s\{2n/n - 1\}/s\{4n/2\} = \sin (\pi/4)/\sin (\pi/4n).$$

With $n = 3$, this shows that a $\{12/2\}$ can be dissected into a $\{6/2\}$. Figure 20.11 is a ten-piece dissection, which can be improved to a nine-piece one as in Fig. 20.12 if turning over is allowed.*

* See Appendix G.

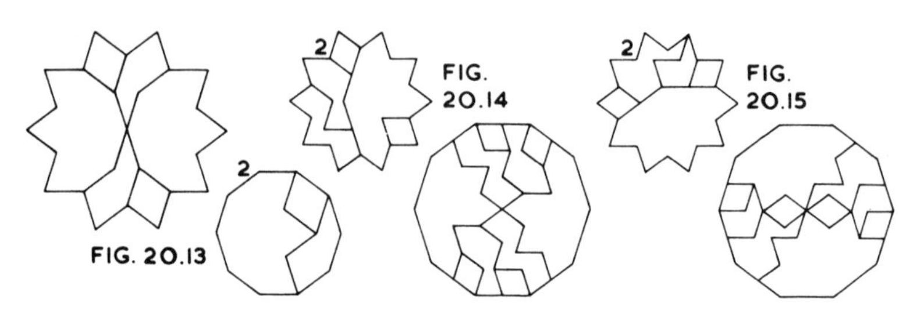

FIG. 20.13

One of the relations revealed by the search is

$$s\{10\}/s\{10/3\} = \sqrt{2}, \tag{8}$$

which indicates dissections of one $\{10/3\}$ into two decagons, of two $\{10/3\}$'s into one decagon, and of $\{10/4\}$ into $\{10/3\}$ because of (6). Collating Figs. 20.7 and 20.13 puts us on the track of the general relation

$$s\{2n\}/s\{2n/\tfrac{1}{2}(n+1)\} = \sqrt{2}, \tag{9}$$

where n is odd. And the decagrams in Fig. 20.16 give a broad hint that leads to

$$s\{2n/\tfrac{1}{2}(n+1)\}/r\{2n/n-1\} = 2 \sin(\pi/2n), \tag{10}$$

where again n is odd. Relations (6) and (9) are two different generalizations of Figs. 20.7 and 20.8. A further RTF is found on dividing (7) by (9), just as we found one by dividing (7) by (6). It is

$$s\{2n/\tfrac{1}{2}(n+1)\}/s\{4n/2\} = 4 \cos(\pi/4) \cos(\pi/4n).$$

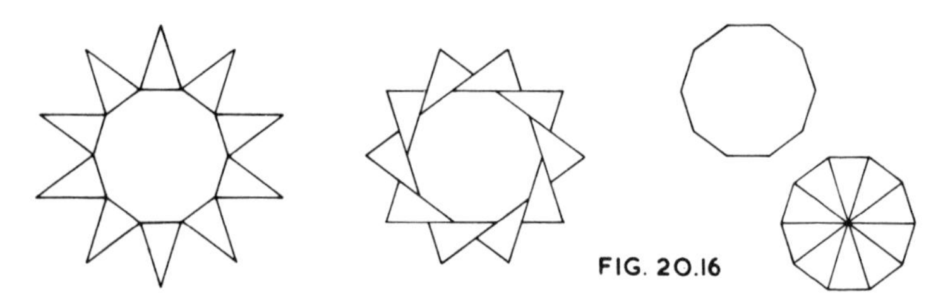
FIG. 20.16

Another RTF we find is

$$s\{5/2\}/s\{10\} = 2 \cos 18°, \tag{11}$$

leading to Fig. 20.17. Multiplying (8) by (11), we get a relation indicating that one {10/3} can be dissected into two {5/2}'s. Having found the simple way of doing this shown in Fig. 20.20, we deduce from it the further relation

$$r\{2n/n - 2\}/s\{n/\tfrac{1}{2}(n - 1)\} = \sqrt{2}, \tag{12}$$

where n is odd. Figures 20.18 and 20.19 show the elementary dissections of a hexagon into two triangles and of a triangle into two hexagons, the rearrangements being omitted; (12) is a generalization.

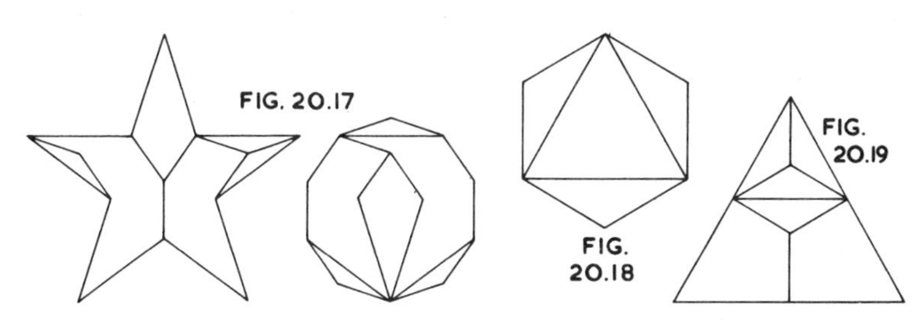

It must be pointed out that the s's in (1) and (3) are not strict analogs; the true analog in a polygram of the s in a polygon is not NO in Fig. 20.20, but NP. I preferred NO because the formulas for s, R, r as in (3) are more symmetric, and because it is usually the more significant dimension in a dissection. With $n = 3$, the right-hand side of (12) is incorrect, but it is quite legitimate to deduce that $r\{6\}/s\{3\}$ is $\sqrt{2}$ times an RTF, for NO in Fig. 20.20 is an RTF of NP and similarly in general.

Substituting $n = 5$ in (10) and (12) shows that a {10/4} too should be dissectable into two {5/2}'s. And, indeed, we find Fig. 20.21.

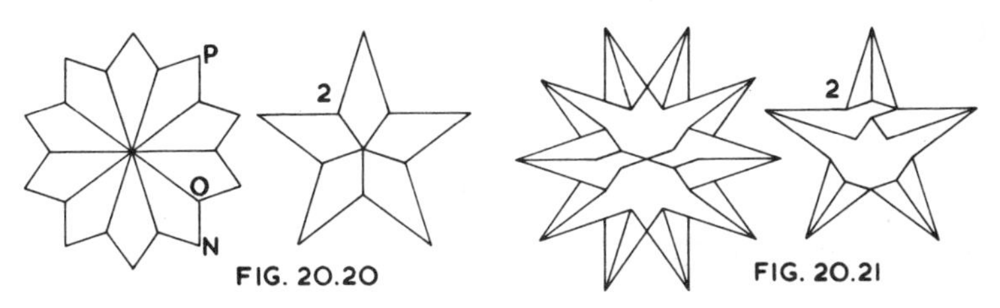

Substituting $n = 7$ in (5), (9), (10), (12), we see that there are dissections of

{14/2} and {7},
{14/4} and two {14}'s,
{14/4} and {14/6},
{14/5} and two {7/3}'s.

Missing members are {14/3} and {7/2}. We guess that this might be because they go together, and, sure enough, we find

$$R\{14/3\}/R\{7/2\} = \sqrt{2} \cdot \cos(2\pi/7),$$

which shows that there are dissections of one {14/3} into two {7/2}'s, and of two into one. (I have not looked for them.) The relation of which this is a particular case is

$$R\{2n/\tfrac{1}{2}(n-1)\}/R\{n/\tfrac{1}{4}(n+1)\} = \sqrt{2} \cdot \cos\{\pi(n+1)/4n\}, \tag{13}$$

where n is one less than a multiple of 4. The least value of n is 3, giving Figs. 20.18 and 20.19 again. Thus, we have in (13) a second generalization of these dissections.

Multiplying (4) by (6), we get

$$s\{4\}/s\{12/5\} = \sqrt{6} = 2\{\tfrac{1}{4}(\sqrt{6}+\sqrt{2}) + \tfrac{1}{4}(\sqrt{6}-\sqrt{2})\}$$
$$= 2\,(\cos 15° + \sin 15°). \tag{14}$$

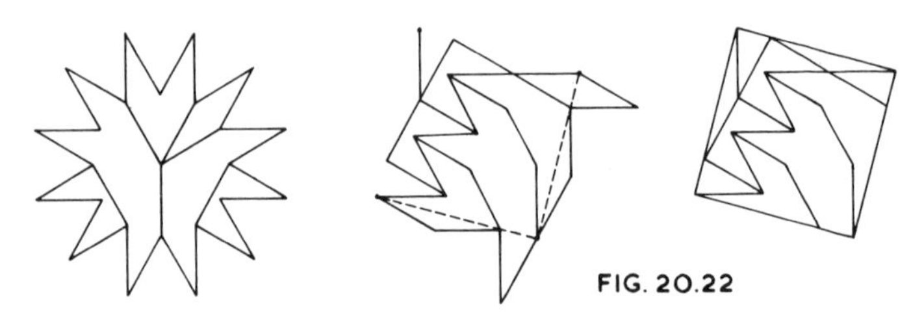

FIG. 20.22

This prepares us for economical dissections of {12/5} into square, and the RTF gives a hint on how to set about finding Figs. 20.22 and 20.23.

Although a dodecagon is easily dissected into a Greek cross as well as into a square, this does not appear to apply to a {12/5}. But we find ample compensation. For the $\sqrt{6}$ in (14) shows that there should be a dissection of a {12/5} into six squares, therefore, into a Latin cross. Figure 20.24 fulfills this expectation.

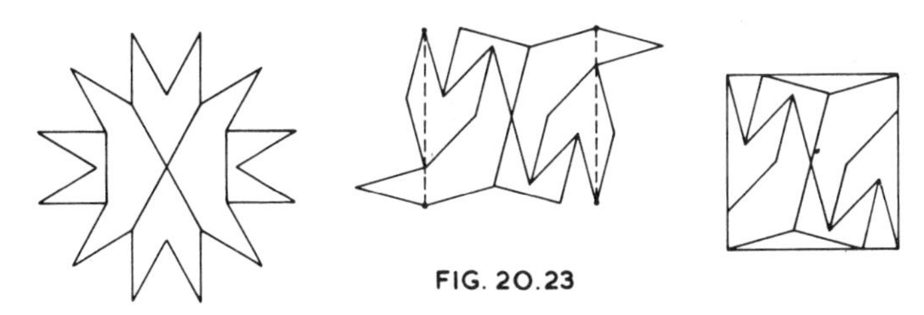

FIG. 20.23

Figures 20.22 and 20.23 are derived from the relation (4) between dodecagon and square, which, so far as I can see, is not a particular case of a general relation. Another apparently isolated relation is

$$s\{3\}/s\{12/2\} = 6 + 4 \cos 30°, \tag{15}$$

leading to Fig. 20.25.*

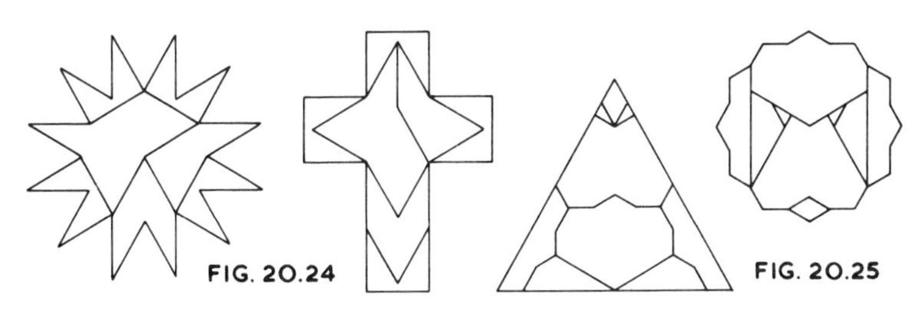

FIG. 20.24 FIG. 20.25

The last example will be used to make clear a possible source of misunderstanding. Dividing (15) by (5) with $n = 6$, we obviously get an RTF. Nevertheless, there is no dissection of one hexagon and one triangle like those in this chapter. This is because the RTF (more easily found from (1) with $n = 3$ and 6) is $\sqrt{2} \cdot \tan 60°$, in which the coefficient is $2 \sin 45°$. But 45° is *not* a multiple of 30° (i.e., π/n). A dissection must, therefore, include cuts whose lengths are not RTF's, and which, therefore, are not determined by intersections of chords. Similarly, there is no dissection of hexagram and hexagon based on RTF's, so the angle between strips at L in Fig. 3.9 is 45°, which is not a multiple of 30°.

It will be appreciated that the theory expounded in this chapter predicts the existence of relatively economical dissections, but gives only the meagerest indication of how to find them. This

* See Appendix G.

can be obvious or obstinate. Thus, dissecting a {10/3} as in Fig. 20.20 was easy, but the {10/4} in Fig. 20.21 was difficult. For all that, there is a fascination in these jigsaw puzzles, in which you have to find the shapes of the pieces as well as where to put them.

Most of the dissections in this chapter were found by seeking pairs of equal dimensions in the respective polygons. The RTF gives a clue to one such pair, and others are easily found by trial, by superposing polygons drawn on separate pieces of tracing paper. In this way you would soon find that the distances *KL* in Fig. 20.5 are equal, and this starts you off with piece *M*. After a while you become familiar with the geometry of the polygons, and wonder at your initial perplexity. A suitable problem for testing one's prentice hand is the assembly of two {8/2}'s or {8/3}'s into one. 12- and 14-piece dissections respectively are found with little trouble, and with perseverance 11- and 12-piece ones, respectively, can be found.

For convenience a list is appended of the general relations found, and of the particular relations of which no generalization could be found.

$$s\{n\}/s\{2n/2\} = 4\cos(\pi/2n),$$
$$R\{2n\}/s\{2n/n-1\} = \sqrt{2},$$
$$s\{2n/n-1\}/s\{4n/2\} = \sin(\pi/4)/\sin(\pi/4n),$$
$$s\{2n\}/s\{2n/\tfrac{1}{2}(n+1)\} = \sqrt{2}\ (2 \mid n+1),$$
$$s\{2n/\tfrac{1}{2}(n+1)\}/r\{2n/n-1\} = 2\sin(\pi/2n)\,(2 \mid n+1),$$
$$s\{2n/\tfrac{1}{2}(n+1)\}/s\{4n/2\} = 4\cos(\pi/4)\cos(\pi/4n)\,(2 \mid n+1),$$
$$s\{2n/n-2\}/s\{n/\tfrac{1}{2}(n-1)\} = \sqrt{2}\,(2 \mid n-1),$$
$$R\{2n/\tfrac{1}{2}(n-1)\}/R\{n/\tfrac{1}{4}(n+1)\} = \sqrt{2}\cos\{\pi(n+1)/4n\}\quad(4 \mid n+1);$$

$$s\{5/2\}/s\{10\} = 2\cos 18°,$$
$$s\{3\}/s\{12/2\} = 6 + 4\cos 30°,$$
$$s\{4\}/R\{12\} = 2\cos 30°,$$
$$s\{4\}s\{12/5\} = 2(\cos 15° + \sin 15°).$$

21

E PLURIBUS UNUM

The title of this chapter indicates that we shall be considering the assembly, by dissection, of two or more polygons into one. It is a more familiar tag than the converse *ex uno plura*. Besides, for uniformity we shall refer to a dissection as that of k polygons into one, not the other way round.

21.1. The simplest assemblies of this kind are of four congruent polygons into a single similar one. No matter what the polygons are, it suffices to draw one or two or three small ones inside the large one, and if necessary add a few lines parallel to sides of the polygon (and usually equal to sides of a small polygon). The irregular hexagon in Fig. 21.1 accommodates three small ones, and in this case no more lines need be added, for the three parallelograms fit round the central piece to make the fourth small hexagon.

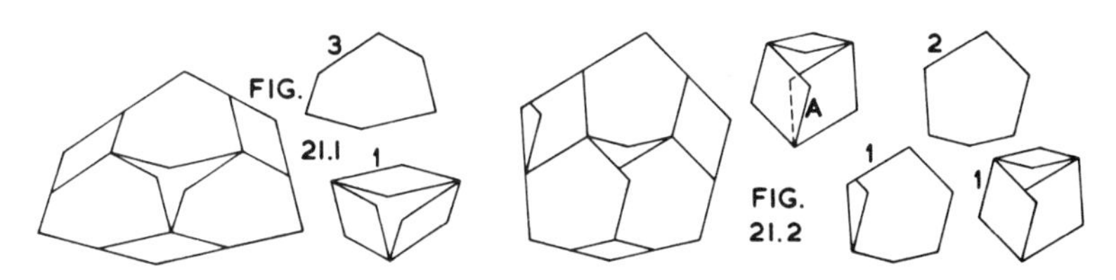

Only two small hexagons have been fitted into the large one in Fig. 21.2. But we draw as much as we can of a third one, and try to assemble the parallelograms and the central piece into a fourth hexagon. We find an overlap as at A. It is got rid of, and the third hexagon completed, by cutting piece A out of either of the overlapping parallelograms. In this or some other way, moderate resourcefulness will always find a means to the end.

The complete small polygons in the large one need not all be

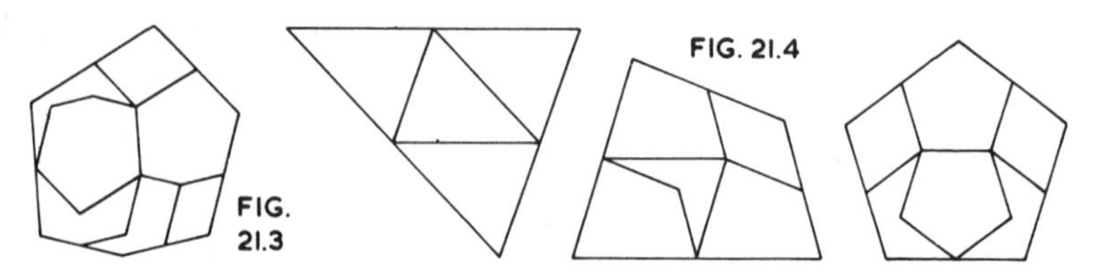

the same way up. In Fig. 21.3, which shows the same hexagon as before, one of the small hexagons is inverted. Nevertheless, cuts can be found to complete the dissection.

It is only human to start by drawing as many small polygons as possible in the large one. But drawing a single one will be found to suffice. An example by C. Dudley Langford is the assembly in Fig. 21.5 of four regular nine-gons into one.

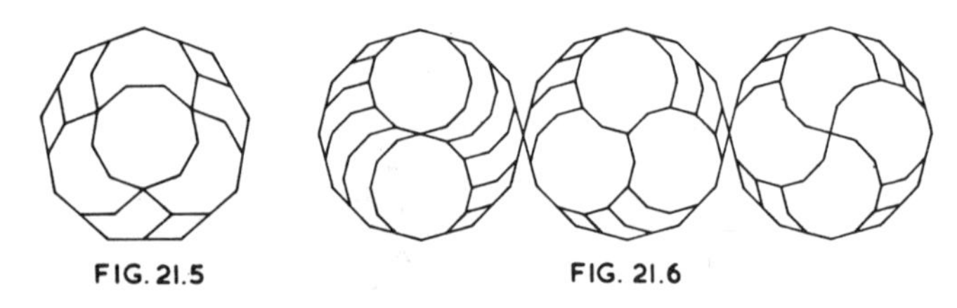

FIG. 21.5 FIG. 21.6

From the examples already given, and Fig. 21.4, one infers that the minimum number of pieces for n-gons in general is $n + 1$. But regular n-gons require only n pieces if n is even. As an example we have in Fig. 21.6 three symmetric assemblies of four dodecagons into one, each assembly requiring 12 pieces. These are referred to, but not described, by H. M. Cundy and C. D. Langford, *Math. Gaz.*, 44, 46 (1960), Note 2875. There have probably been several independent discoveries of them.

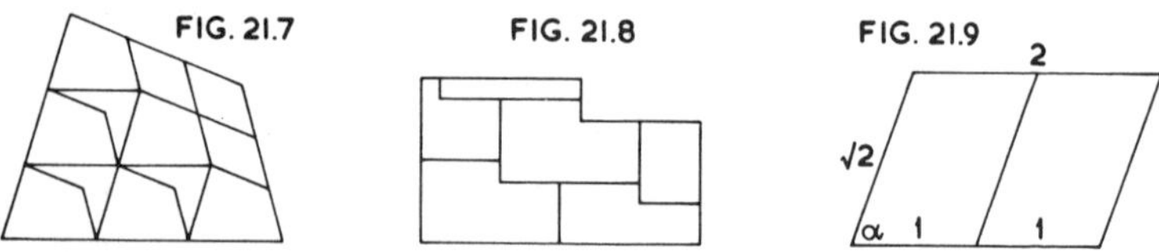

It is just as easy to assemble nine small polygons, or indeed any square number of them. Let Fig. 21.7 suffice.

If the polygon is a tessellation element, an economical assembly may be found by superposition. Figure 21.8 is an example in which the polygon is that of Fig. 9.17. But if the polygon has to be cut to make it a tessellation element, superposition is likely to be inferior unless the polygon has a large number of sides. For instance four regular pentagons can be assembled into one in six pieces; but the number of pieces to make them tessellation elements is already eight.

21.2. We go on to the assembly of a nonsquare number of polygons into one, dealing first with methods suitable only for equal polygons. If the polygon has a tessellation with a period parallelogram whose sides have a ratio $\sqrt{k} : 1$, there is an assembly of k polygons into one. For as Fig. 21.9 shows when $k = 2$, a parallelogram compounded of k period parallelograms, turned over if α is not a right angle, is similar to the period parallelogram. Hence, we need only superpose. The octagon tessellation in Fig. B8 and the hexagram one in Fig. B14 are like this, with $k = 2$ and 3 respectively. But in neither case do we get dissections as good as can be found by trial.

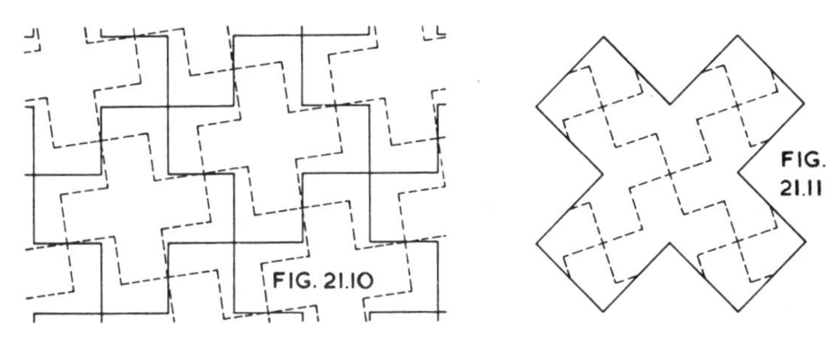
FIG. 21.10
FIG. 21.11

If the polygon has a square tessellation, we can find dissections in which the number of small pieces is $x^2 + y^2$. For if we superpose the tessellations of large and small polygons at an angle whose tangent is y/x or x/y, then points congruent in the former coincide with points congruent in the latter. A simple example is Dudeney's dissection of two Greek crosses and one. In Fig. 21.10, congruent lines in the two tessellations are at 45° as required, but the crosses themselves are not, because one of the tessellations has been turned over. To illustrate a higher value of k, an assembly of five crosses into one is shown in Fig. 21.11. Here $x = 2$, $y = 1$, so the angle between congruent lines is to be $\tan^{-1} \frac{1}{2}$, and again a better dissection is obtained by turning one of the tessellations over.

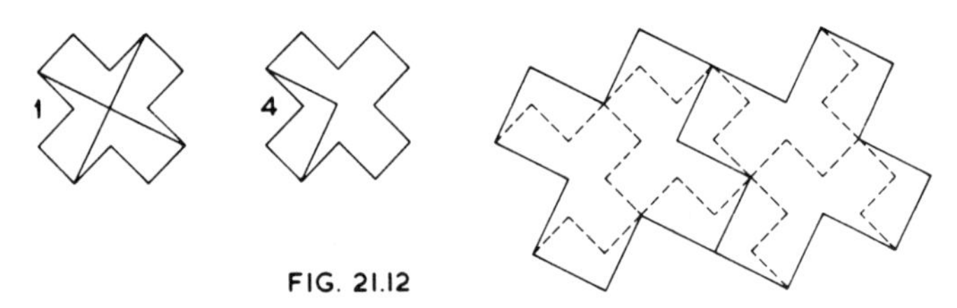

FIG. 21.12

The assembly of k polygons with a square tessellation into h, where $h = p^2 + q^2$, is effected similarly. For k into one, the angle between congruent lines is $\tan^{-1}(y/x)$; for h into one it is $\tan^{-1}(q/p)$ or $\tan^{-1}(p/q)$; therefore, for k into h it is either

$$\tan^{-1}(y/x) - \tan^{-1}(q/p) \text{ or } \tan^{-1}(y/x) - \tan^{-1}(p/q).$$

Figure 21.12 shows five Greek crosses assembled into two, the angle between congruent lines (and tessellation elements) being

$$\tan^{-1} 2 - \tan^{-1} 1 = \tan^{-1} 1/3.$$

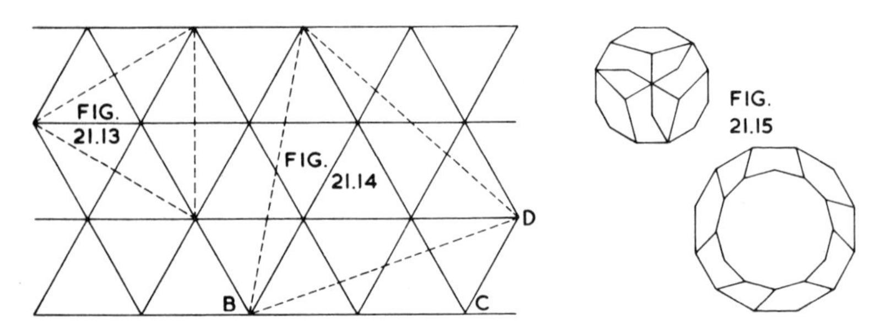

FIG. 21.13

FIG. 21.14

FIG. 21.15

A similar procedure applies to tessellations in which congruent points are the vertices of equilateral triangles. Let the lattice points in Figs. 21.13 and 21.14 be congruent points in a tessellation. The cosine formula applied to a triangle such as $\triangle BCD$ shows that the square of the distance such as BD between any two lattice points (with CD as unit of length) is of the form $p^2 + pq + q^2$, and all integers of this form can be expressed as $x^2 + 3y^2$. Hence, if k has the latter form, tessellations of large and small polygons can be superposed so that points congruent in the former coincide with points congruent in the latter. Assemblies of three triangles and of seven into one are shown in Figs. 21.13 and 21.14. The assemblies of three hexagons into one, seven into one, seven into three, and so on, discussed in Chapter 11, are also of this type.

It should not be overlooked that in the expressions $x^2 + y^2$ and $x^2 + 3y^2$, y can be 0; that is, the procedure described in this section covers assemblies of a square number of polygons into one, or into another square number.

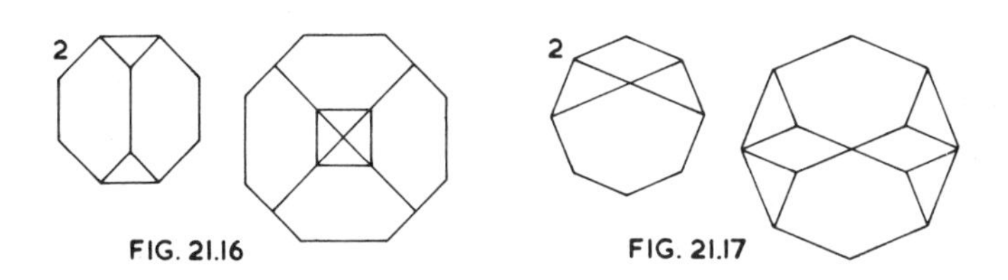

FIG. 21.16 FIG. 21.17

21.3. Methods based on tessellations do not always give good results (though they may be suggestive of better). Nor do they apply to all simple assemblies of k polygons into one. From (1) and (3) in Chapter 20 we can deduce for what regular polygons and polygrams there should be simple dissections of this kind. The principle is the same as in deducing the existence of simple star dissections. Thus, two n-gons can be assembled into one if $\sqrt{2}$ is an RTF of π/n, and so if n is a multiple of 4. This was discovered by Freese. Figures 21.15-21.17 confirm the cases $n = 12$ and $n = 8$. We also expect assemblies of $2p^2$ polygons as in Fig. 21.18, which shows an octagon dissected into one eighth of a larger one.

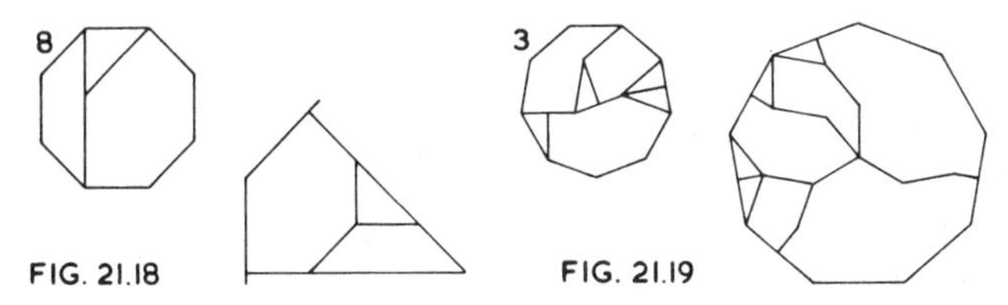

FIG. 21.18 FIG. 21.19

Likewise, there are assemblies of three polygons into one if $\sqrt{3}$ is an RTF of π/n, and so if n is a multiple of 3. This was discovered by Langford, whose assembly of three nine-gons into one is given in Fig. 2.19. (This figure is made clearer by showing only one dissected small nine-gon in the large one.) The first cases $n = 3$ and $n = 6$ are illustrated in Figs. 21.13 and 11.2. Of the assemblies with $n = 12$ in Figs. 21.20 and 21.21, the first is more elegant, but not minimal.

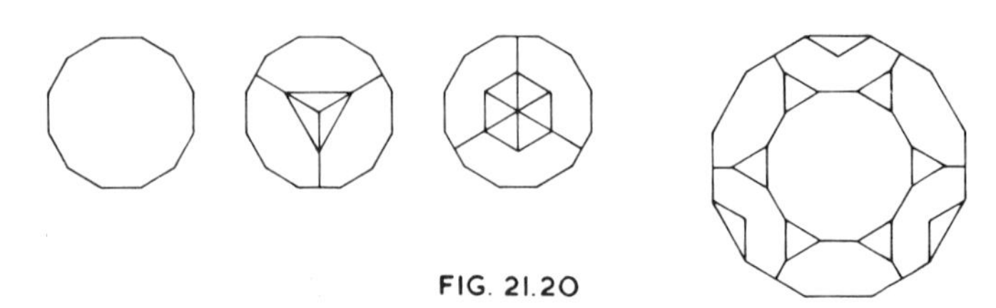

FIG. 21.20

The foregoing applies equally to polygrams, and, as there are more of these than of polygons, plenty of gaps remain for the ambitious dissector to fill in. Once again the hexagram proves amenable, as we see from the assembly of three into one in Fig. 21.22. The {12/3} is not much harder, whatever you may think at the first sight of Fig. 21.23. A few moments' study will show how the assembly of two {12/3}'s into one is just an elaboration of that of two squares into one. But one of the elaborations is that four pieces G are to be assembled into two pieces H. These pieces are elements of square tessellations, so one gets the required assembly of four into two by superposition, as shown enlarged on the right. Altogether there are 16 pieces.

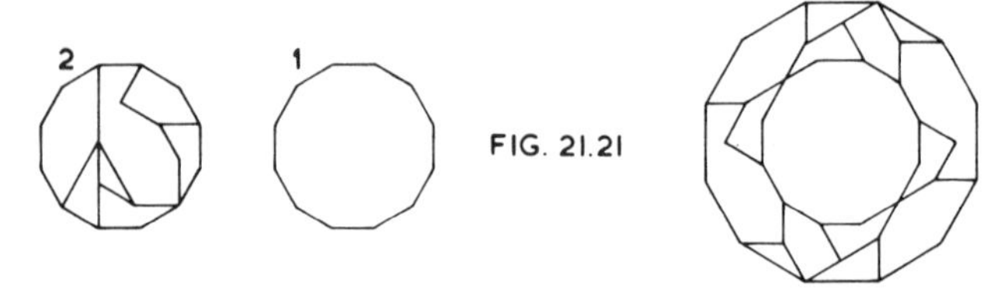

FIG. 21.21

Some assistance is afforded in finding dissections as just described, if the large and small polygons are both divided into rhombs and triangles whose sides are equal to those of a small polygon. These rhombs and triangles (with a few obvious extra cuts in some cases) give a dissection, and it can be made more economical by combining groups of them into single pieces. Figure 21.19 was found thus.

The radical $\sqrt{5}$ is equal to $4 \cos (\pi/5) - 1$ and $4 \sin (\pi/10) + 1$. Hence, there should be assemblies of five pentagons and of five decagons into one. But we know there are, for Langford has already discovered Figs. 21.24 and 21.25. In the latter, only one fifth of the decagon is shown. Further roots of primes known

FIG. 21.22

to be RTF's are those of 17, 257, 65537. The assemblies of seventeen 17-gons and so forth into one are left as exercises for the reader.

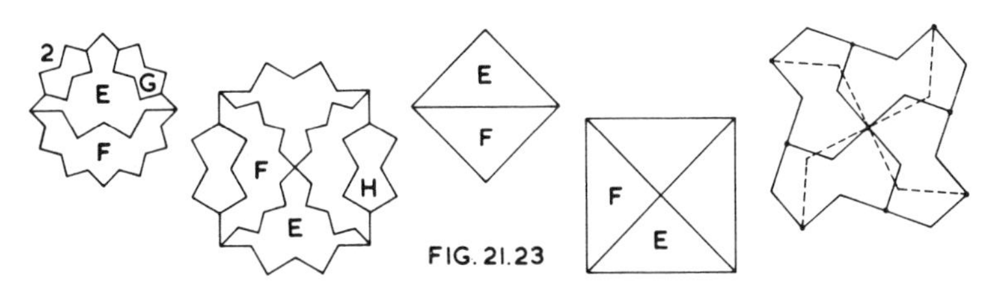

FIG. 21.23

Another radical that is an RTF is $\sqrt{6}$; see (14) in Chapter 20. This indicates an assembly of six dodecagons into one, which, however, I have not bothered to look for. Since $\sqrt{6} = 2\sqrt{(3/2)}$, there should also be an assembly of three dodecagons into two. This time I bestirred myself, and found Fig. 21.26. Here the cuts are oriented so as to bring out the feature that the pieces can be rearranged by translation without rotation. Several of the dissections in Chapters 20 and 21 have this feature.

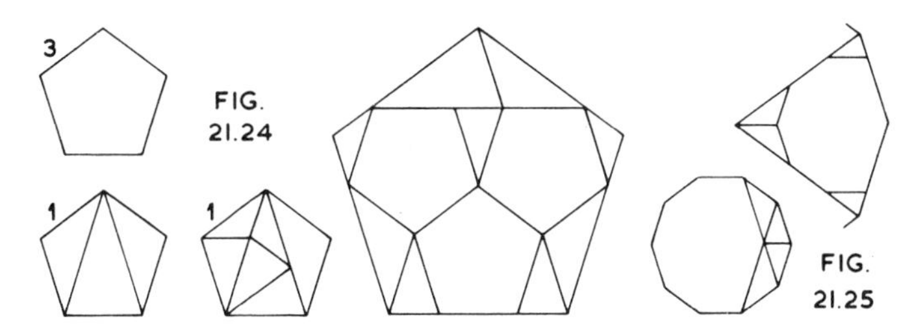

FIG. 21.24

FIG. 21.25

There will be no assembly based on RTF's of two n-gons into one if n is odd or singly even, for $\sqrt{2}$ is not then an RTF. Nevertheless, there is a simple and beautiful assembly of two equal regular n-gons into one, for any value of n greater than 3. This "ring expansion," shown for a pentagon in Fig. 21.27, is one of Freese's enviable achievements. The construction is quite simple: make triangles J isosceles and right angled. It is easily seen that for n-gons there are $2n + 1$ pieces.

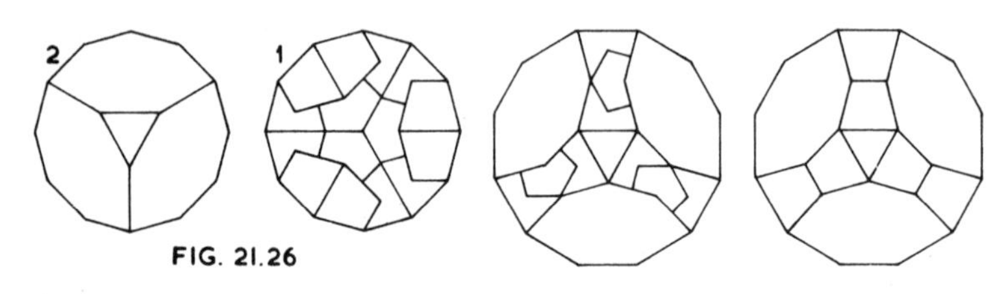

FIG. 21.26

21.4. The assembly of two similar but unequal polygons into one similar one is a more difficult problem. The simplest cases of triangle and square have already appeared in Figs. 1.9, 1.11, and 9.17. For further regular polygons we have Freese's masterly dissection shown for a pentagon in Fig. 21.28. If the sides of the polygons are s_1, s_2, s with $s_1 \leqslant s_2 < s$, then the cuts in the polygon of side s_2 are determined, starting from the center K, by the relations

$$LM = \tfrac{1}{2}s_2, \quad MN = \tfrac{1}{2}s_1, \quad \text{whence } LN = \tfrac{1}{2}s.$$

We can also make the cuts in the smallest polygon, provided only that it is not so small that N is above K. Like the assembly in Fig. 21.27, this one also requires $2n + 1$ pieces for n-gons.

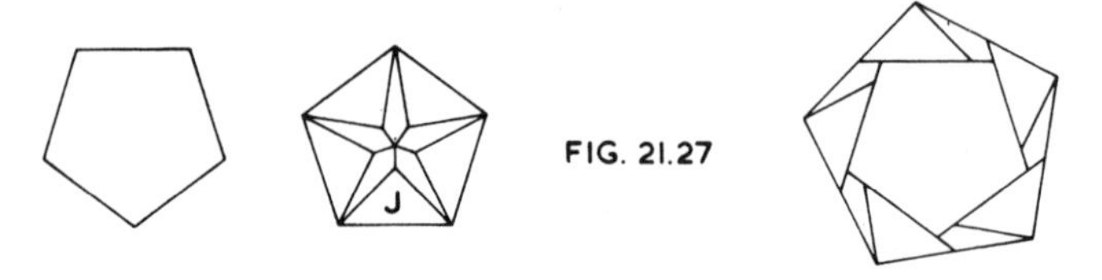

FIG. 21.27

Nothing so elegant as this seems possible if the polygons are irregular. But at least it can be done for any polygon by repeated application of the Q-slide. Thus, the slide can be applied in duplicate to quadrilaterals as in Fig. 21.29. It requires eight pieces as against five for triangles, each extra side in the polygons meaning three more pieces.

The method of Fig. 9.17 can be generalized to rectangles. Two similar rectangles, combined as in Fig. 21.30, form the element

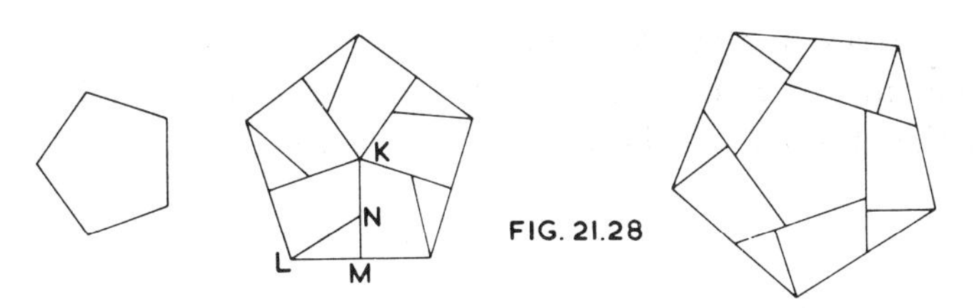

FIG. 21.28

of a tessellation with period parallelogram $OPQR$. If the dimensions of the rectangles are $a \times b$ and $c \times d$ with $a/b = c/d$, then triangles OPS and ORT are similar, from which it easily follows that $\angle POR$ is a right angle. We also have

$$OR^2/OP^2 = (a^2 + c^2)/(b^2 + d^2) = a^2/b^2,$$

that is, the period parallelogram is similar to the two rectangles. Hence, we get an assembly of two rectangles into one, all similar, by superposing a tessellation or strip of large rectangles on the tessellation at the proper angle.

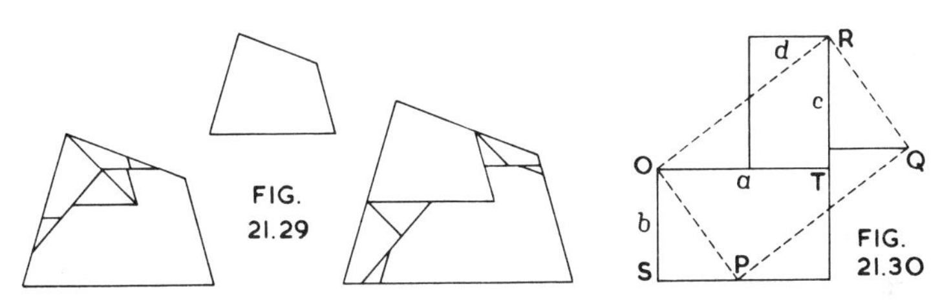

FIG. 21.29 FIG. 21.30

Figure 21.30 shows two golden rectangles assembled into one. To deal with polygons other than rectangles, they are dissected into similar rectangles and a tessellation is drawn, like the one shown in full lines in Fig. 21.31. On it we superpose a tessella-

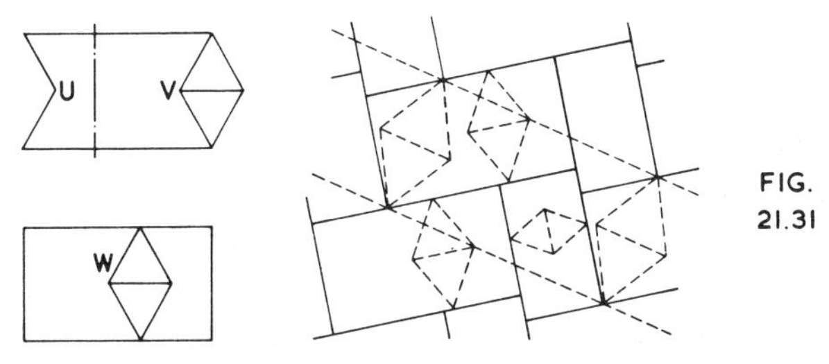

FIG. 21.31

tion or strip derived from the large polygon. (Here, as elsewhere, it suffices to use a single element.) This procedure will be illustrated with hexagrams.

A vertical cut passing anywhere between U and V in Fig. 21.31 changes the hexagram into a rectangle. Draw the tessellation with element compounded of the two rectangles, omitting for the moment the lozenges W, and draw a strip element of the large hexagram. Superpose it on the tessellation and look out as usual for a favorable location, but bear in mind that positions will have to be found for the omitted lozenges. The location in Fig. 21.31 appears to be best, permitting the addition of the lozenges where shown. One of them is cut, but this cannot be remedied without introducing more cuts elsewhere.

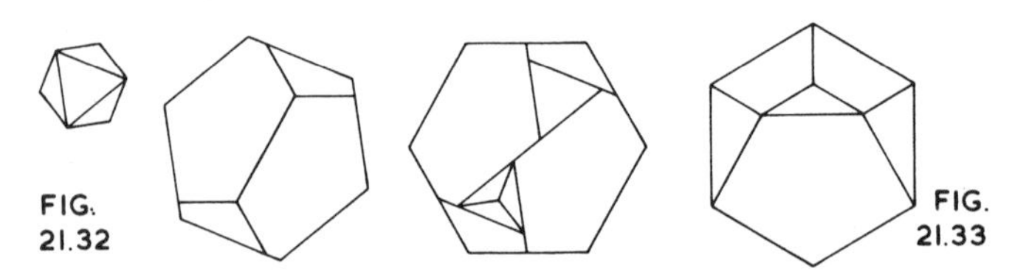

FIG. 21.32 FIG. 21.33

The number of pieces in a dissection like this tends to be rather large. This is only to be expected, for it is a tall order to assemble two polygons into one, if there is no metric relation between the two to take advantage of. But regular hexagons form one of the exceptions, thanks to the tessellations of Figs. 9.18-9.20. We can use them to get more economical dissections than those given by the general method of Fig. 21.28. To do so, we dissect the smaller hexagon into two triangles and get eight- and nine-piece assemblies as in Figs. 21.32 and 21.34.

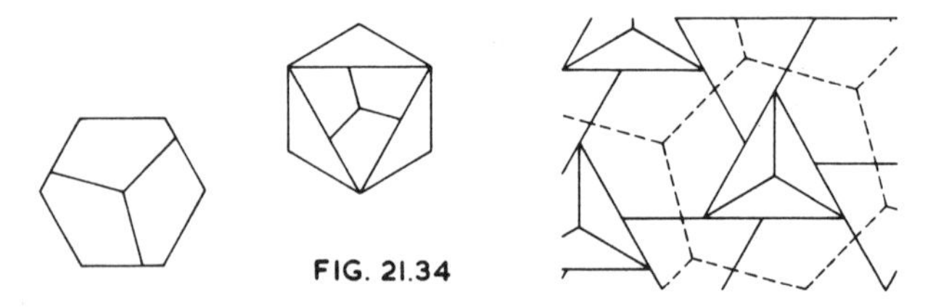

FIG. 21.34

With a favorable metric relation between the polygons, a piece or two can be saved. If the dissection of Fig. 21.32 is applied to hexagons whose sides have the ratio $\sqrt{3} : 1$, it will be found that the tessellation is that of Fig. 9.19 and only seven pieces are required; it is an easy step from the assembly thus obtained to the six-piece symmetric one in Fig. 21.33. (This is another dissection that can be made in the form of a hinged model.) Similarly, if the sides of two triangles have the above ratio, their assembly requires only four pieces instead of five as in Figs. 1.9 and 1.11. This special case is included in Fig. 22.10 (*G* into *H* plus *K*).

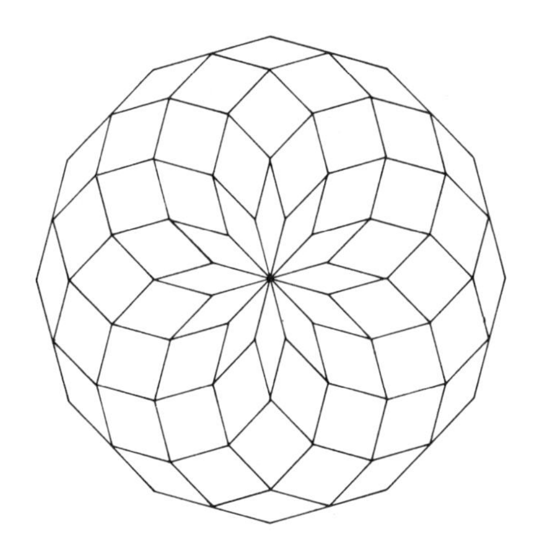

22

THREE-FIGURE DISSECTIONS

The number of combinations of three polygons is legion, but most dissections of three polygons require too many pieces to be interesting. Again, there are two three-figure dissections that are quite economical, but they are too easy and obvious (now that you know all the tricks). These are the dissections of Greek cross, octagon, and square, and of Greek cross, dodecagon, and square. The first is obtained in nine pieces by combining Figs. 12.1 and 22.1, and the second in ten pieces by combining Figs. 9.9 and 22.1. Even though the first is incidentally a minimal dissection of Greek cross and octagon, to illustrate them would be wasteful and ridiculous excess.

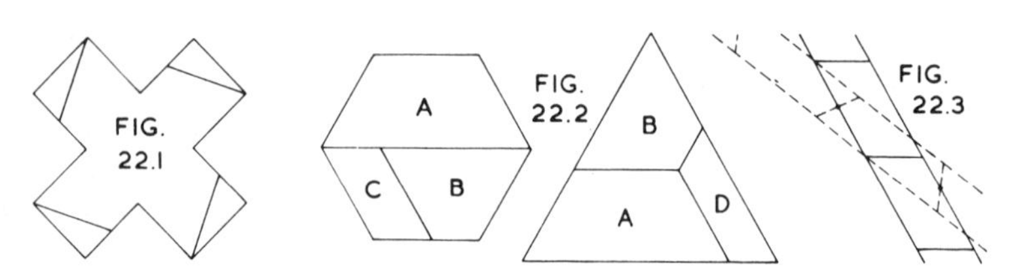

FIG. 22.1

FIG. 22.2

FIG. 22.3

It is different when the polygons are the hexagon, square, and triangle. An early dissection of hexagon and triangle, superseded by Fig. D15, was found to be just the thing for this three-figure dissection. It takes advantage of the very simple relation between the angles in the two polygons, in that we cut out two big pieces *A* and *B* in the hexagon as in Fig. 22.2, with which to fill up as much as possible of the triangle. The parallelogram remnant *C* in the hexagon is then dissected by *PT*, as in Fig. 22.3, into the trapezoid remnant *D* in the triangle. Although this dissection requires six pieces instead of the minimal five, it gives a more economical three-figure dissection because the superposition in

Fig. 22.4 is free from cuts that cross. The nine pieces required all appear in the common area in this figure.

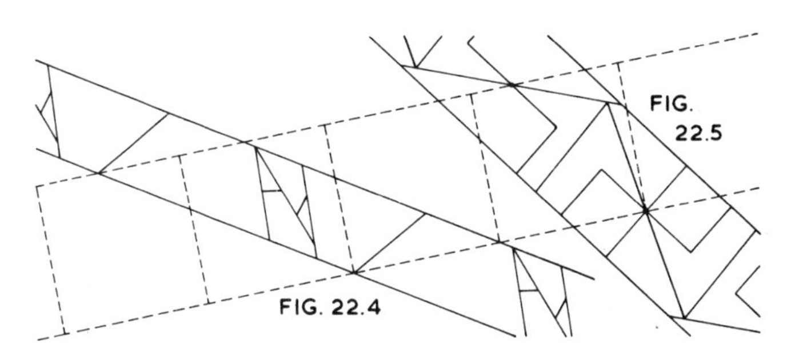
FIG. 22.4
FIG. 22.5

Another example of repeated superposition is provided by the Latin cross, square, and triangle. The *TT*2 dissection of cross and triangle in Fig. 22.5 is copied from Fig. 5.3, and on it we superpose *PT*-wise a strip of squares.

The Greek-cross strip with element as in Fig. 2.18, reproduced in Figs. 22.6 and 22.7, is sometimes useful. If there is a minimal or near-minimal dissection using this strip, then adding a vertical line *EF* gives a further dissection into a square. The fewer the lines crossed by *EF* the better. Trying this with the dissection of cross and hexagon in Fig. 3.7, we find that the added cut *EF* unavoidably crosses two existing ones. But this is where the unbeautiful alternative there referred to comes into its own. As is seen from Fig. 22.6, the added cut, shown chain-dotted, crosses only one existing cut.

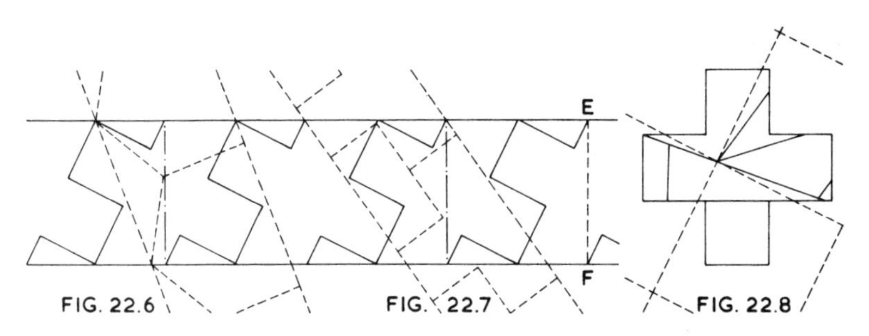

FIG. 22.6
FIG. 22.7
FIG. 22.8

An example using a nonminimal dissection is that of Greek and Latin crosses and square. The dissection of the crosses in Fig. 22.7 is not as good as Fig. 4.11, there being eight pieces instead of seven. But the additional cut to get the square crosses only one existing cut in Fig. 22.7.

If two of the polygons are the Greek cross and square, an economical dissection will sometimes be found if a tessellation of squares is superposed on the Greek cross complete with the cuts that dissect the latter into the third polygon. One example of this is the dissection of cross and pentagon, taken from Fig. 5.8 and reproduced in Fig. 22.8. Trial soon shows that the best superposition is the one illustrated.

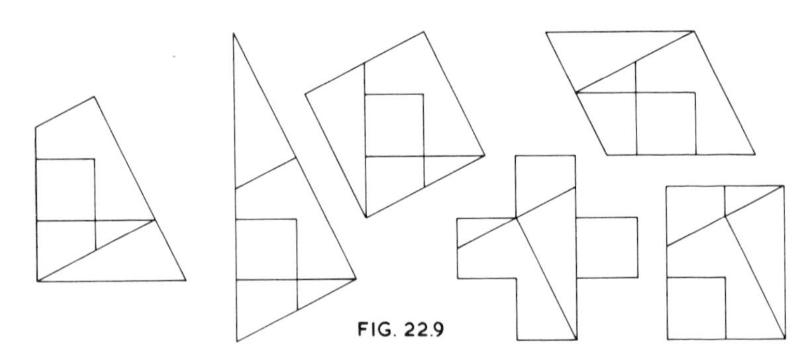
FIG. 22.9

The three-figure dissections described are, of course, only the easiest ones. There must be others awaiting discovery that do not require an excessive number of pieces. The search for them is complicated by the fact that a minimal three-figure dissection need not be based on minimal two-figure dissections.

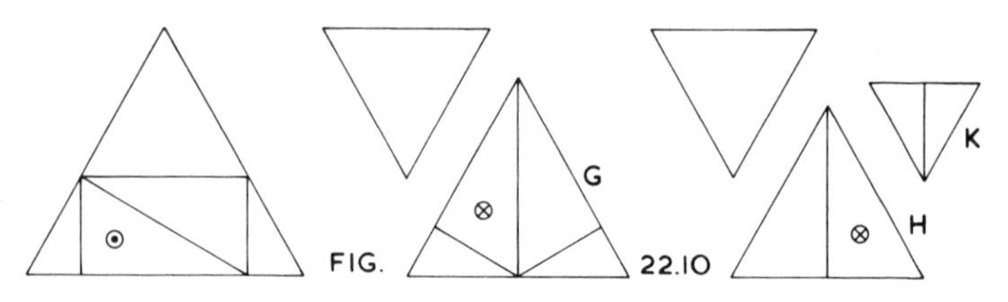

FIG. 22.10

There are at least two dissections that fall outside the treatment in this chapter. In the tangrammatic Fig. 22.9, due to Loyd, we have for good measure not three polygons but six. And Fig. 22.10 shows a dissection by Dudeney of one triangle into two and into three.

23

CURVILINEAR DISSECTIONS

The other chapters in this book, all concerned with rectilinear dissections, can be read in deep seriousness. But this hardly applies to curvilinear dissections, whose nature fits them only for puzzles and more or less artistic designs.

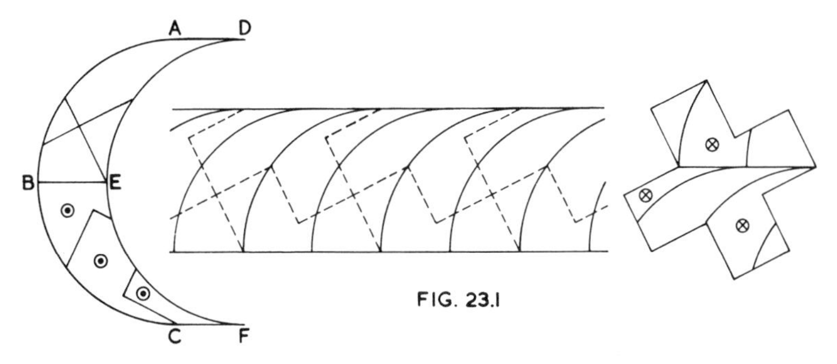

FIG. 23.1

The difficulty in finding curvilinear dissections is that to every convex edge of a figure there must nearly always correspond somewhere else concave edges of the same length and radius for it to fit into. For instance, in the dissection of crescent and

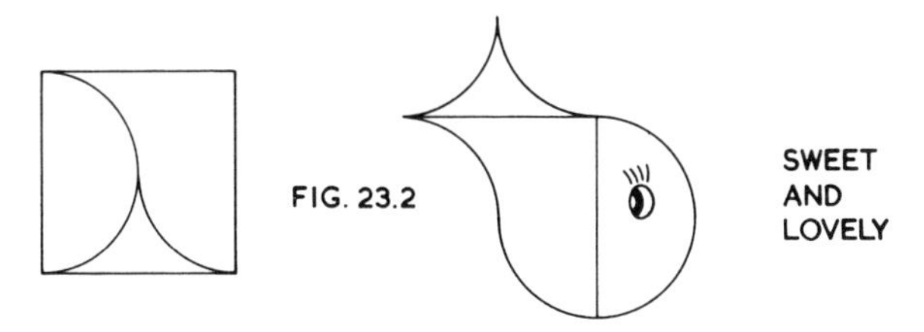

FIG. 23.2

Greek cross in Fig. 23.1 the edges of the former must include two equal curved portions *ABC* and *DEF,* and the remainder must be two straight lines *AD* and *CF* (if the figure is to look anything like a crescent).

Again, the shapely contour of the mermaid's rival in Fig. 23.2 has equal lengths of convex and concave curves. But Loyd exploited the difficulty by the trick shown in Fig. 23.3.

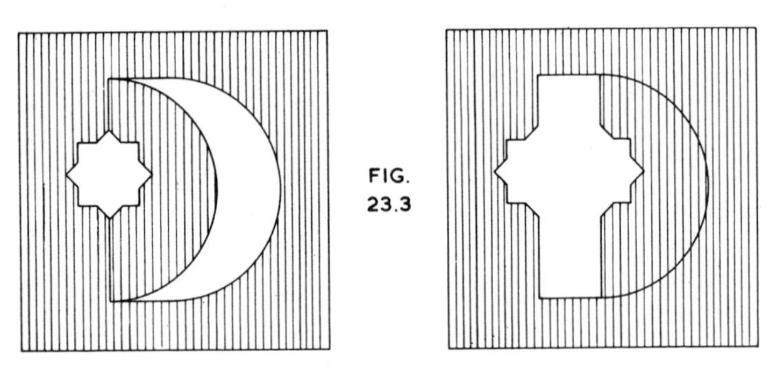

FIG. 23.3

Because of this difficulty, one cannot as a rule choose two curvilinear figures beforehand and try to dissect them; it is rather a matter of seeing what a given curvilinear figure can

FIG. 23.4

be dissected into. Thus it is with the circle. One might have in mind its dissection into some kind of curvilinear star, but hardly a goal more precise. And you count yourself lucky to find something as passable as Figs. 23.4-23.6.

FIG. 23.5

THE STARFISH

The radius of the cuts in Fig. 23.4 is arbitrary. There is less freedom in Figs. 23.5 and 23.6, in that all cuts must have the same radius as the circle.

Working out the cuts for Fig. 23.4 is quite easy. Draw curves of convenient radius joining the vertices G, H, J, K of an n-gon, making L such that GK and KL are equal and collinear; join KL by a curve of the same radius, and complete by drawing a circle through the n points like L. This construction differs from that of Fig. 19.7 only in that the lines are curved.

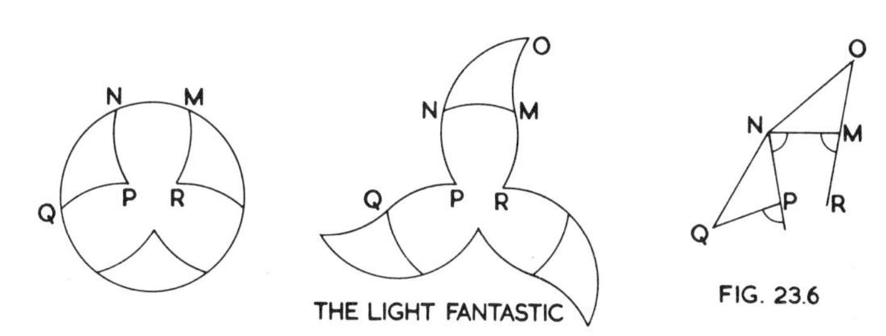

FIG. 23.6

But Figs. 23.5 and 23.6 require a more elaborate approach, to find lines that give smooth joins at points M in the star. (Points N give no trouble, for in curvilinear angles

$$\angle ONM + \angle MNP = \angle NQP + \angle MNP = \angle QNP + \angle MNP = \pi.)$$

To see one's way more clearly in considering the join at M, replace some of the curves in Fig. 23.6 by straight lines as on the right. Then a smooth join at M requires that O, M, R be collinear, whence the three marked angles must be equal. For an n-point star, since arc MNQ is $1/n$ of the circumference, it follows that

$$\angle MNP = \tfrac{2}{3}\angle MNQ = \tfrac{2}{3}(\pi - \pi/n).$$

Since also

$$QP = PN = NM = MO = MR$$

and their length can be given any convenient value, all the points named in Fig. 23.6 can be determined, and, therefore, all points in the star dissection, and the radius of the arcs is that of the circle through M, N, Q. In going about it this way, the radius of the circle is whatever it turns out to be. If on the contrary we start with a circle of given radius R, then we use the fact that

$$MN = 2R \sin \theta,$$

where

$$\cot \theta = \cot (\pi/n) + 2 \sin (\pi/3n + \pi/6)/\sin (\pi/n).$$

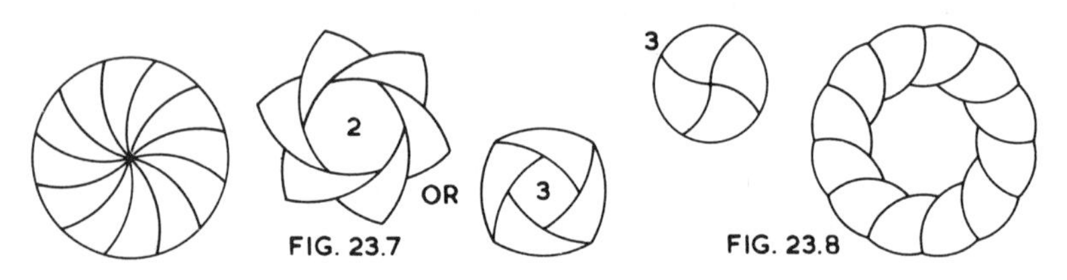

FIG. 23.7 FIG. 23.8

One might also have in mind the dissection of a circle into some kind of ring. Figures 23.7 and 23.8 are typical of what can be found. There are numerous variants, depending on the number of sectors per circle and the number of rings per circle or circles per ring.

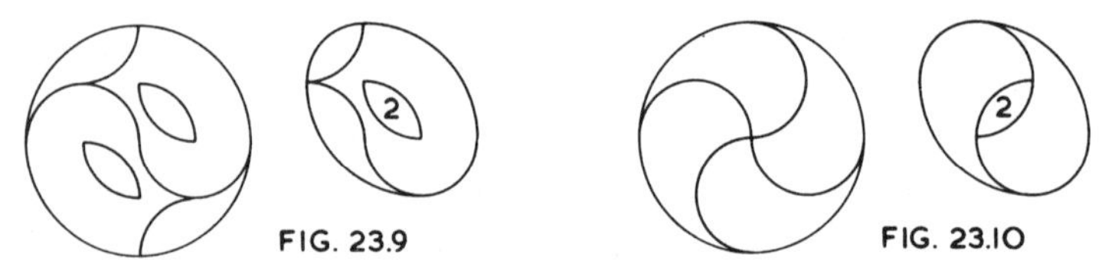

FIG. 23.9 FIG. 23.10

Somewhat similar is the second of the dissections of a circle into two oval stool seats with hand-hole. (Both the dissections are by Sam Loyd—*ex ungue leonem.*) With the orientation of Figs. 23.9 and 23.10, all centers for drawing arcs are at lattice points on a grid of squares of side $\frac{1}{2}R$, where R is the radius of the circle, and their radii are either R or $\frac{1}{2}R$.

FIG. 23.11

The horseshoe dissection in Fig. 23.11 leans heavily on Fig. 23.9. But the spade-heart dissection in Fig. 23.12 is an isolated curiosity, and an outstanding one.

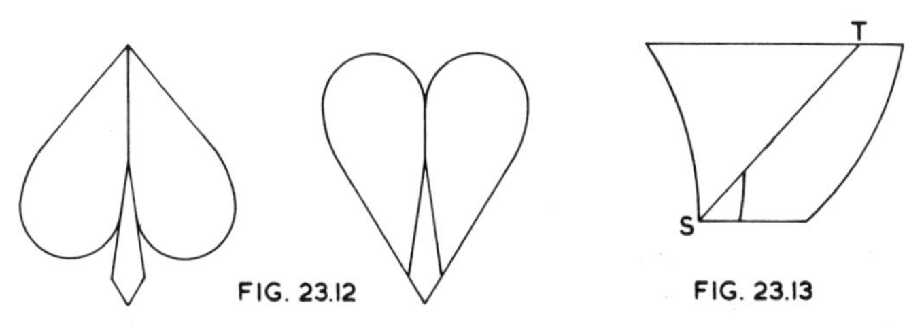

FIG. 23.12

FIG. 23.13

Figure 23.1 shows that an *S*-dissection can be applied to curvilinear figures. So can the *P*-slide, as we see from Fig. 23.13; here the transverse cut *ST* also can be curved. Perhaps a good dissection based on this can be found.

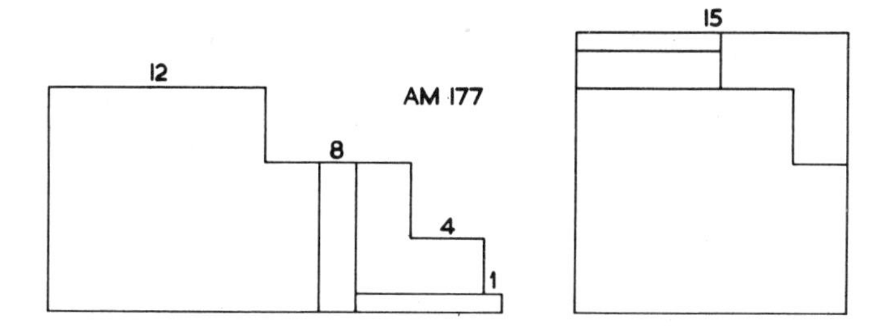

AM 177

24

SOLID DISSECTIONS

To prove that any polygon can be dissected into any other of the same area, one shows that any polygon can be cut up into triangles, and that each triangle can be dissected into a rectangle whose length is equal to the side of the square equivalent to the polygon. Stacking the rectangles gives a dissection of polygon and square, so we have the successive dissections: first polygon → square → second polygon.

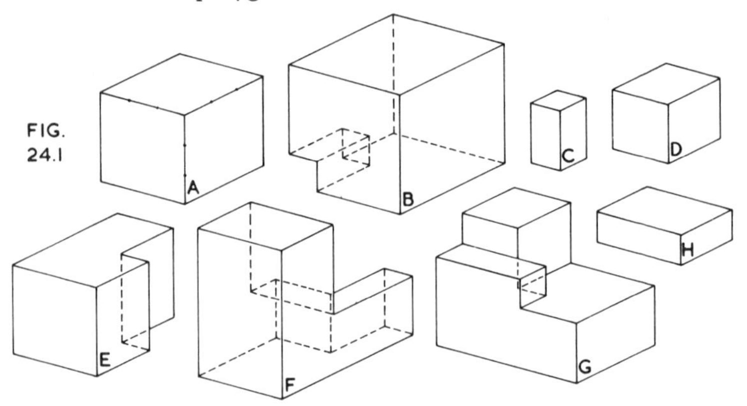

FIG. 24.1

In three dimensions the triangles would be replaced by pyramids (or tetrahedrons), and the square by a cube. Now it is true that any solid bounded by plane surfaces can be cut up into pyramids, but it is not true in general that a pyramid can be dissected into a cube or a slice of a cube. The best we can do is to replace the triangles by triangular prisms with parallel end-faces. If each of the two solids can be cut up into such prisms, then one of them can be dissected into the other. But plenty of solids can not, so the analog of the two-dimensional proof does not get us very far.

The consequence is that our choice of solids for dissection is somewhat limited. Moreover solid dissections tend to require a large number of pieces and are hard to visualize, so it is not surprising that very little has been done in this field. Attention has been confined mainly to showing that certain solid dissections are possible, the finding of economical ones being passed over. The first two dissections to be described are exceptions.

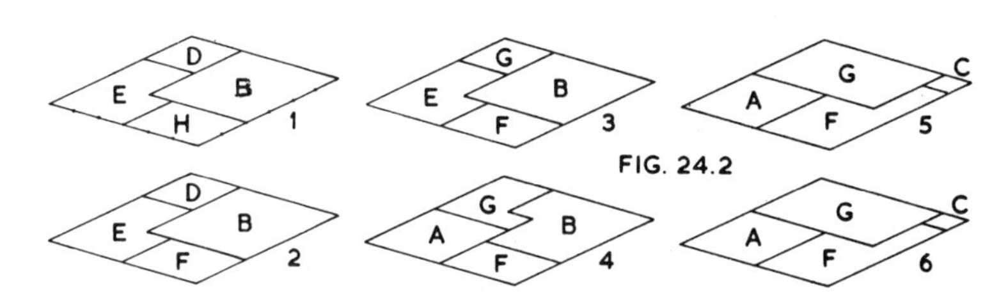

FIG. 24.2

24.1. A number of solid dissections are based on methods that are counterparts of methods used in plane dissections. For instance there is an eight-piece *R*-dissection, due to R. F. Wheeler, of a cube of side 6 units into three cubes of sides 3, 4, and 5 units; it provides a concrete demonstration of the numerical identity

$$3^3 + 4^3 + 5^3 = 6^3.$$

FIG. 24.3

All pieces in the dissection are assemblies of the 216 unit cubes into which the large cube can be divided. The unit of length is marked by the dots on piece *A* in Fig. 24.1, which shows the separate pieces, while Fig. 24.2 is a key to their positions in the cube of side 6. The six drawings therein show which pieces appear in six successive layers of unit thickness, starting from the top. The dissection into the three cubes consists of piece *A*, pieces *B* and *C*, and pieces *D-H*. In the last of these the orientations of *E* and *F* are different from those shown in Fig. 24.1, so the new orientations are included in Fig. 24.3.

I wonder how many hours of patient trial went into the working out of this dissection. My compliments to its originator.

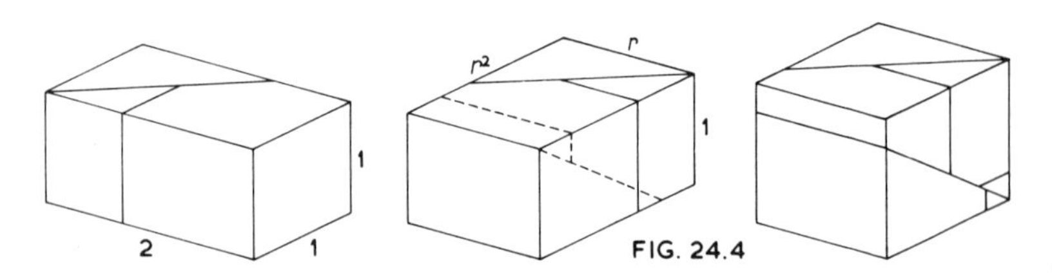

FIG. 24.4

24.2. Some of the simpler solid dissections are in effect combinations of plane dissections, each of one face, carried down in depth to a second parallel face. Such is A. H. Wheeler's dissection of a $2 \times 1 \times 1$ rectangular block into a cube. Figure 24.4 shows a *P*-slide on the top face, locating vertical cuts through the block which change its dimensions to $r^2 \times r \times 1$, where r denotes the cube root of 2. A second *P*-slide, shown in broken lines, changes the $r^2 \times 1$ faces to $r \times r$, and locates cuts along planes perpendicular to these faces which dissect the block into a cube. After the first *P*-slide there are three pieces; the long cut in the second *P* slide cuts each of them, making six; and the short cut adds one more piece, making seven in all.

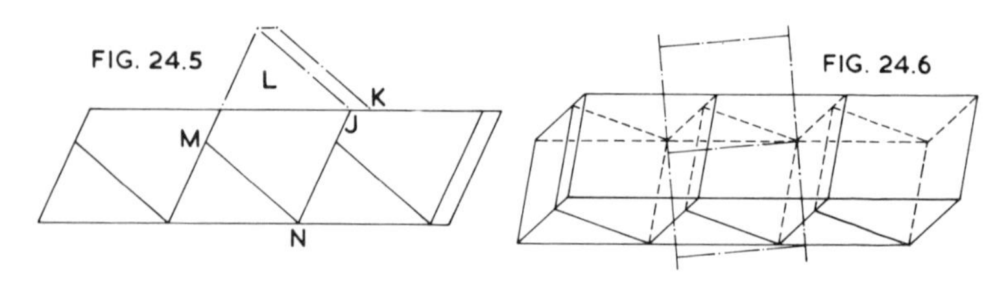

FIG. 24.5 FIG. 24.6

24.3. The general dissection of one parallelepiped into another can be effected by means of solid strips that are "superposed" (actually they interpenetrate). The cuts in the common volume give the dissection. But the strips—imagined to lie horizontally —must have the same height. Therefore, one of the parallelepipeds must be modified, for instance as in Fig. 24.5, which is a side elevation of a strip showing only visible lines. In this figure a cut has been made through edge *JK*, and piece *L* has been placed on face *MN*. Figure 24.6 is a plan view of the same strip, crossed by a second, interpenetrating strip. The latter is shown chain-dotted because in this figure ordinary broken lines have their usual significance in engineering drawing. The second strip is one made up of rectangular blocks, to simplify the drawing, but this special case adequately illustrates the general case. The preliminary cut in the first block makes two pieces; each is cut by a vertical face of the second strip, making four; and the plane common to abutting elements in the second strip cuts three of the four, making seven pieces in all.

This method does not give a minimal dissection of $2 \times 1 \times 1$ block and cube. Whether we start by making the block the same height as the cube or vice versa, in the superposition the plane common to abutting elements in the second strip cuts all four existing pieces in the first strip-element instead of just three.

The foregoing general method uses solid *P*-strips. Solid *T*-strips exist also, but are only of limited use. For trial will show that the conditions for their existence and for the existence of dissections based on them are most stringent; e.g., certain faces must be perpendicular, or certain lengths or angles in the respective solids must be equal. The stringency precludes a dissection of any generality, so examples using *T*-strips will not be described.

If you have no objection to turning over in plane dissections (implying a third dimension to turn over in), you may have none to the analogous feat of turning inside out in solid dissections (implying a fourth dimension). The restrictions will then be somewhat eased.

It is also possible to use honeycombs (the three-dimensional equivalent of tessellations). For instance the solid equivalent of a Greek cross, made up of seven cubes, is the element of a honeycomb with a period parallelepiped. Two suitable pairs of congruent planes will demarcate a strip; or the solid cross can be dissected directly into other solids having the same period parallelepiped.

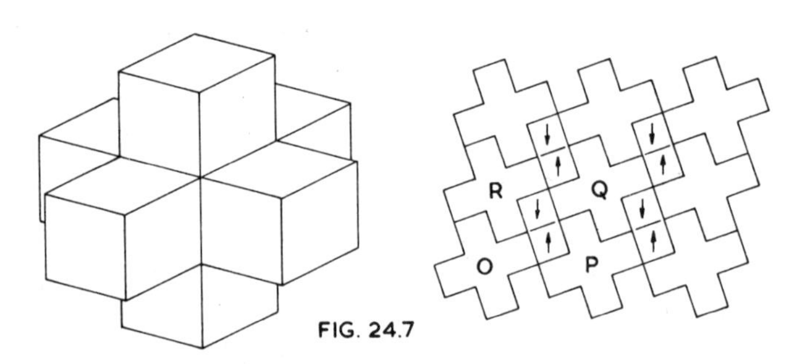

FIG. 24.7

In the honeycomb of crosses each layer is as in Fig. 24.7. The crosses shown contribute cubes to the layers above and below with period parallelogram *OPQR*. The spaces marked with upward arrows can be filled by cubes sticking up from the layer below, and those marked with downward arrows, by cubes sticking down from the layer above.

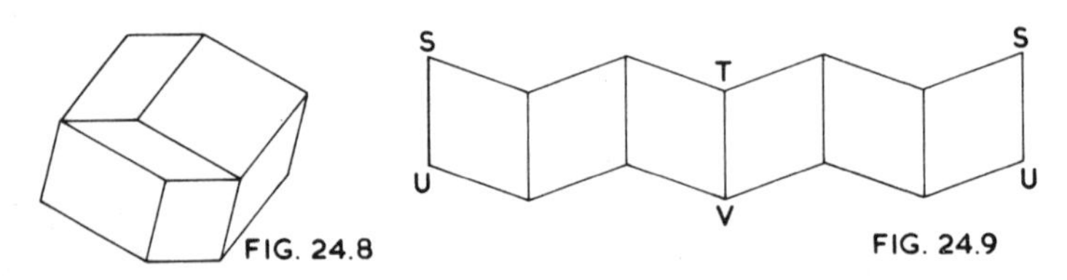

FIG. 24.8

FIG. 24.9

24.4. The most general class of solids known to be dissectable into a cube, and, therefore, into one another, is the class of *zonohedrons.* A zonohedron is a convex solid, every face of which has opposite sides equal and parallel, and is, therefore, dissectable into parallelograms; see for instance the octagonal face divided as in Fig. 24.8. If every face not already a parallelogram is thus divided, we can regard each parallelogram as a separate face of the zonohedron and say that *all* its faces are parallelograms. The simplest zonohedron is the parallelepiped. Next comes the flat solid (a slice of a hexagonal prism) whose upper and lower faces are parallel-sided hexagons, the respective vertices of which are joined in pairs by six equal parallel lines, not necessarily perpendicular to the upper and lower faces. It is easily realized that any set of faces joined by parallel edges, such as the set shown spread out in Fig. 24.9, cannot spiral away from its first member and, therefore, joins up with it to form a closed zone (Greek for *belt)* encircling the zonohedron. Hence the name, and hence the dissection, as we shall see.

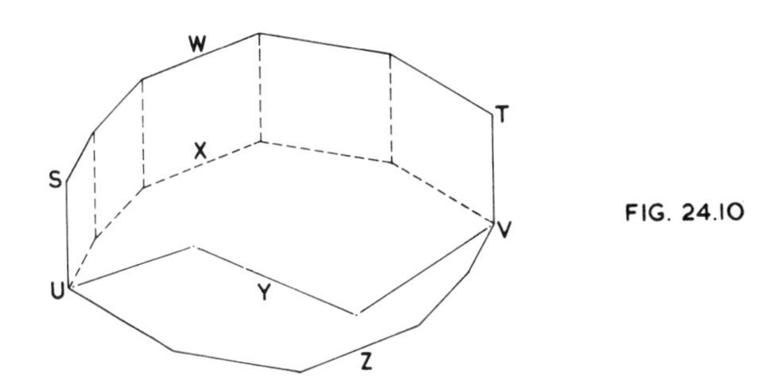

FIG. 24.10

It will be shown that a zonohedron can be dissected into parallelepipeds. Figure 24.10 is meant to be only the silhouette of a zonohedron, so that all you see of a more or less horizontal zone is its edges *SU* and *TV*. (Its lower front edge might be like *UYV*. Its upper or lower edges do not in general coincide with *SWT* or *UZV*.) Remove from the zonohedron a layer bounded above by the top surface *SWT* and below by an internal surface *UXV* whose vertical separation from *SWT* is everywhere the same, namely, the constant vertical distance *SU* or *TV* in Figs. 24.9 and 24.10. The layer removed can be divided into parallelepipeds, their upper faces being faces of the zonohedron, each lower face being identical with the upper face above it, and respective vertices of these faces being joined in pairs by equal and parallel lines. Hence, we have dissected part of the zonohedron into parallelepipeds, and we are left with a zonohedron with fewer faces, for those forming the zone have been removed. We can similarly remove layer after layer until the zonohedron we are left with (if any) is just one parallelepiped. The final steps are to dissect the parallelepipeds into slices of the cube and stack them. Although a zonohedron is defined as a convex solid, re-entrant ones can be included, e.g., the layer *SWTVXU* removed from Fig. 24.10.

This dissection was first explicitly described by Leo Moser. It does not embrace all the possible dissections into a cube. But it is moderately comprehensive for all that, being the most general solid dissection to date, and probably the most general one possible.

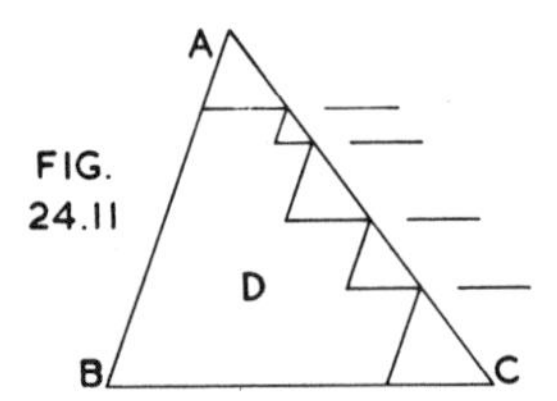

FIG. 24.11

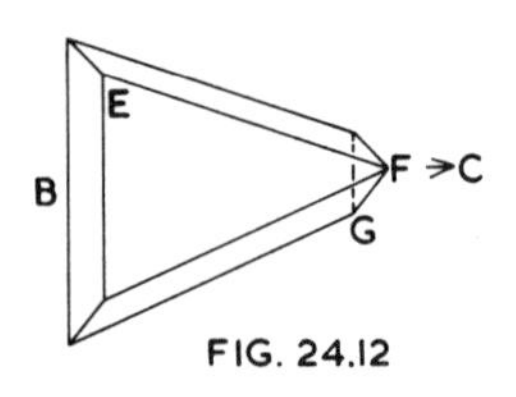

FIG. 24.12

24.5. Only one other kind of solid dissection, which can lay claim to generality, is known to me. It can be applied to any polyhedron, and will be described with respect to a tetrahedron. Figure 24.11 is a side elevation of a tetrahedron, oriented so that *AB* and *BC* represent faces and *AC* an edge. The small triangles along *AC* represent tetrahedrons of the same shape as *ABC*. If they are removed, the solid that is left can be dissected into a cube, as was proved by J. P. Sydler. The proof consists in dividing the tetrahedron into slices by horizontal cuts as indicated by the lines on the right of *AC*. The lowest slice, shown in Fig. 24.12 as viewed from above, can be divided into two triangular prisms with parallel end-faces by a cut through plane *EFG;* similarly with the other slices. And the prisms obtained from all the slices can be dissected into a cube.

Making $AC = a$ and dividing a in any way into any number k of equal or unequal parts $a_1, a_2, \cdots, a_k$, we can say that the tetrahedron of dimension a can be dissected into k tetrahedrons of dimensions $a_1, a_2, \cdots, a_k$, plus a cube. Since any polyhedron can be divided into tetrahedrons, the generalization from tetrahedron to any polyhedron is immediate.

Figure 24.14 shows the pyramid obtained by joining an upper vertex *H* of the cube in Fig. 24.13 to the lower vertices *J, K, L, M,* the cube and pyramid being viewed from above with the eye to the southwest. By virtue of the previous paragraph, this pyramid can be dissected into three equal smaller ones plus a cube. But three pyramids like that in Fig. 24.14 can be assembled into a cube, their three vertices meeting at *H,* and the square bases of the other two pyramids being erected on *KL* and *LM*. It follows that such a pyramid can be dissected into two cubes, and so into one.

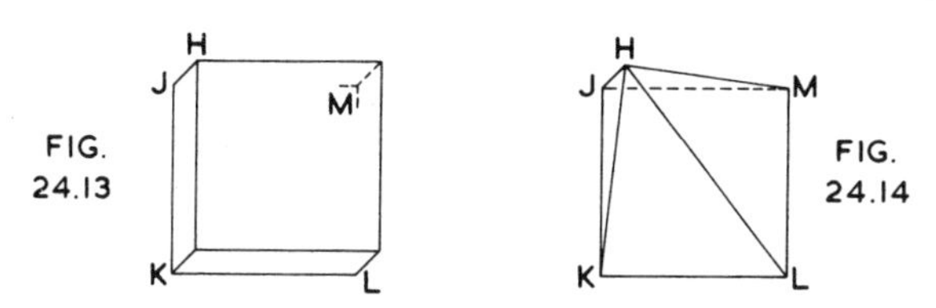

FIG. 24.13

FIG. 24.14

Four pyramids as in Fig. 24.14 can be assembled with their *HJ* edges coincident to form what is known as a Juel pyramid. Its apex, therefore, lies over the center of the square base, and its height is half the side of the base. Clearly a Juel pyramid also can be dissected into a cube.

There are *n*-dimensional dissections for all values of *n*. When *n* is even (as in Chapters 1-22), any figure can be dissected into any other. The figure can be that which corresponds to a triangle and tetrahedron. This implies the existence of *PT* and *TT* dissections that are thoroughgoing generalizations of the procedure to 2*m* dimensions, free from the excessive restrictions on three-dimensional dissections. Who will find them?

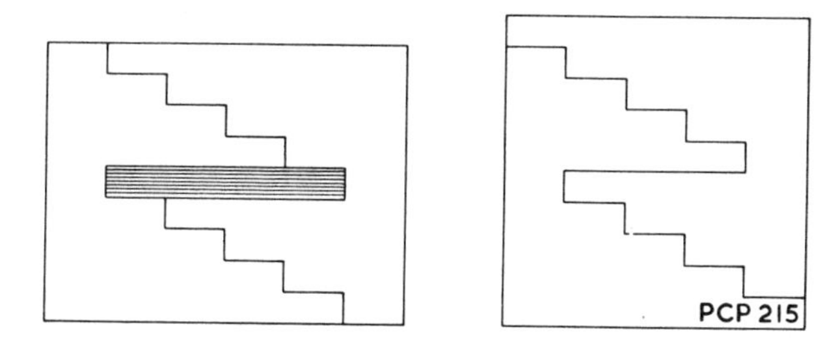

POSTSCRIPT

It is probable that if a paper on dissections were submitted to a first-class mathematical journal, the editor would turn up his nose at it. For, unlike crossing bridges over the Pregel or combing a hairy sphere, dissections are not regarded as serious mathematics. They are useless, because they have no apparent application to any other problem.

But here is an application. If n points are distributed in a unit square and joined by a continuous network, then its length can always be made not greater than some limiting value l_n, no matter how the n points are distributed (*Math. Gaz.*, 44, 182 (1960)). It is known that

$$l_n = k\sqrt{n} + O(1) \ (k \leqslant 1).$$

What if the n points are in a unit equilateral triangle? A little reflection will show that, by means of dissections, the same formula can be proved for the triangle, or indeed for any unit polygon.

Does the skeptic object that the application too is useless? The retort is that the uselessness of the application doesn't matter—the subject is worth pursuing if it has *any* application. Gradually a chain of successive applications accumulates, all of them as useless as you like. And one fine day, one of the applications will be found to be really useful, galvanizing the whole chain.

The subject will then acquire a dignified name such as ?-ology, denoting the study of ?-omorphic transformations. Post-graduate students will do research in it, and thereby get a Ph.D. The subject will then have gained added usefulness and luster in that *man kann damit promovieren,* as Landau said of number theory.

Meanwhile, the solution of dissection problems depends on the stimulus of their intrinsic interest, and not on the lure of a doctorate. The humblest and most obvious problem is to whittle down the numbers of pieces required for individual dissections. This can depend on patient trial; thus the seven-piece dissection of V and square in Fig. 18.16 was the outcome of far more experiment than the text indicates. But it can also depend on a modicum of inspiration; one example is the discovery that the pieces of dodecagon in the tessellation of Fig. B12 have the other arrangement in Fig. B13.

On a higher plane is the problem of turning polygons into tessellation elements. Finding *S*-dissections has been pushed back into finding strips, and finding strips (in the more difficult cases) has been pushed back into finding tessellations. What can finding tessellations be pushed back into?

Other teasers (so it seems) are to develop some kind of theory of turning over, and of dissections that can be made as hinged models. Chapters on both topics were originally projected, but my findings were too scrappy to be worth devoting a chapter to.

Perhaps the most interesting problems are associated with the subject-matter of Chapters 20 and 21. One of them is to devise a method of finding these dissections that depends less on ingenuity and more on routine. At present we have only the assistance afforded by the RTF and the method of dividing into rhombs and triangles explained with reference to Fig. 21.18. Neither of these gets us very far; contrast this with the methods using strips and tessellations. The nature of the method awaiting discovery is at present veiled in mystery, but once found it may seem as obvious as strips and and tessellations do now.

Another problem associated with Chapters 20 and 21 is to discover further general relations such as (5) and (6) in Chapter 20. If you try this problem, say by seeking generalizations of the particular relations at the end of Chapter 20, you will probably be surprised at your helplessness in handling those sines and cosines you thought you were so familiar with.

For those not satisfied with two dimensions, there is a virtually untouched field in solid dissections. Here the problem may be to find something worth doing, as well as to do it. But such things as a minimal dissection of rhombic dodecahedron and cube may be worth trying. Finally, if you are greedy for more dimensions, try the following problem in four: Describe the general *PT* dissection of parallelotope and pentatope, and the general *TT* dissections of two pentatopes.

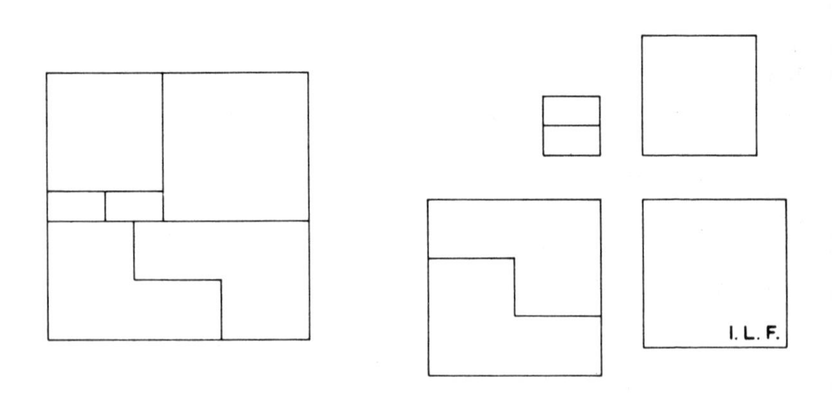

Appendix A

PROBLEMS FOR SOLUTION

You are invited to try your strength on the problems that follow, before looking up the solutions in Appendix D. Problem 3 requires that you dissect a hexagon using the same strip as in Fig. 2.9, into a golden rectangle using the same strip as in Fig. 2.23; it is a *PP* dissection; it can be done in five pieces. The rest of the problems are stated similarly.

1. Rectangle 1.2, square; *P;* 4.
2. Miter 5.9, square; *P;* 5.
3. Hexagon 2.9, golden rectangle 2.23; *PP;* 5.
4. Hexagram 2.17, square; *PP;* 5.
5. Hexagram 2.17, golden rectangle 2.23; *PP;* 5.
6. Hexagram 2.17, octagon 2.14; *PP;* 9
7. Greek cross 2.20, golden rectangle 2.23; *PP;* 5.
8. Latin cross 2.3, golden rectangle 2.23; *PP;* 5.
9. Latin cross 2.3, octagon 2.14; *PP;* 8.
10. Latin cross 2.3, hexagram 2.17; *PP;* 9.
11. Open out a cube (see text to Fig. 1.7) to get a figure with a four-piece dissection into a square.
12. Pentagon 4.1, golden rectangle 2.23; *PT;* 6.
13. Hexagram 2.17, hexagon 4.4; *PT;* 7.
14. Golden rectangle wide strip, triangle 4.3; *TT*2; 4.
15. Hexagon 4.4, triangle 4.3; *TT*2; 5.
16. Hexagon 4.4, pentagon 4.1; *TT*2; 7.
17. Hexagon 4.5, pentagon 5.5; *TT*2; 7.
18. Octagon 2.15, triangle 4.3; *TT*2; 8.
19. Octagon 2.15, pentagon 5.5; *TT*2; 9.
20. Greek cross 4.6, triangle 4.3; *TT*2; 5.
21. Dodecagon 9.10, triangle 4.3; *PT;* 8.

22. Dodecagon 9.10, golden rectangle 2.23; *PP;* 7.
23. Maltese cross B16, Greek cross B15; *T;* 9.
24. Swastika B16, Greek cross B15; *T;* 8.
25. Octagon 2.10, hexagon parallel to *AB* in 7.1; *PP;* 9.
26. Hexagram 2.17, dodecagon parallel to *AB* in B12; *PP;* 10.
27. Heptagon 15.2, square; *TT*2; 9.
28. Heptagon 15.6, golden rectangle 2.23; *PT;* 9.
29. Heptagon 15.2, hexagon 4.5; *TT*2; 11.
30. Greek cross 4.6, heptagon 15.2; *TT*2; 12.
31. Nine-gon 16.9, square; *TT*2; 12.
32. Nine-gon 16.8, golden rectangle 2.23; *PT;* 12.
33. Decagon 17.1, pentagon 4.1; *PT;* 11.
34. Decagon 17.2, golden rectangle wide strip; *PP;* 8.
35. Decagon 17.4, hexagon 2.8; *PP;* 9.
36. Latin cross 2.3, decagon 17.8; *PP;* 10.
37. Decagon 17.2, octagon 2.14; *PP;* 13.
38. Greek cross 2.18, decagon 17.8; *PP;* 11.

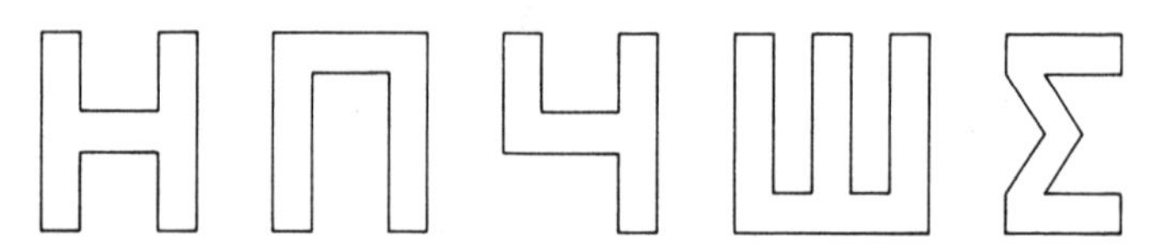

39. H or pi as above, square; *P;* 7.
40. Russian cha, square; *T;* 5.
41. Russian sha, square; *T;* 6.
42. Sigma, square; *P;* 5 (with turning over).
43. {8/3}, square; *P;* 8.
44. {8/3}, square; completing tessellation; 9 (not minimal).
45. {8/2}, Greek cross; *T;* 8.
46. Assemble two {8/2}'s into one; *U;* 11.
47. Assemble two {8/3}'s into one; *U;* 12.
48. Assemble an {8/2} and an {8/3} of same *s* into an {8/3}; *U;* 9.
49. Assemble a {12/3} and a {12/4} of same *s* into a {12/5}; *U;* 13.
50. Assemble four heptagons into one; *U;* 8.
51. Assemble four pentagrams into one; *U;* 14.

52. Assemble five pentagrams into one; *U;* 20.

53. Find a nine-piece dissection of Greek cross, square, and triangle.

54. With three straight cuts, divide a $24 \times 9 \times 8$ rectangular block into four pieces that can be assembled into a cube.

55. Divide a cube of side 6 by cuts parallel to faces into six pieces that can be assembled into six different rectangular blocks.

56. Divide the pyramid of Fig. 24.14 into six pieces that can be assembled into a $3 \times 3 \times 1$ rectangular block.

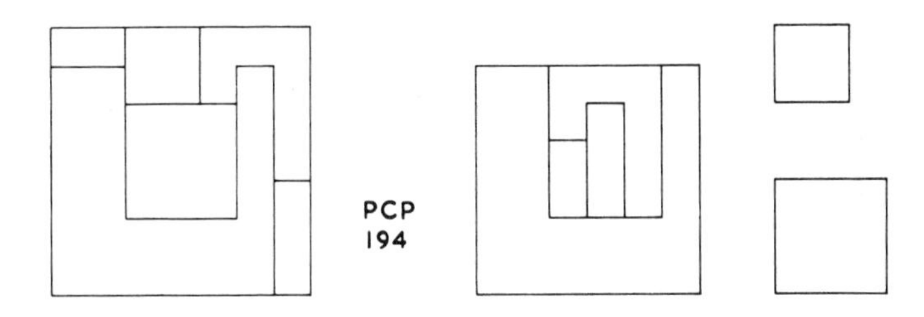

Appendix B

DRAWINGS OF POLYGONS, ETC.

This Appendix caters to those who would like to experiment with strips and tessellations, but shy at preparing drawings from the data in Appendix C. Instead, they need only make tracings from the drawings that follow. Most of the strips used in this book can be obtained from them.

The polygrams have areas 8 and 16 sq. cm. Otherwise the area is 16 sq. cm. throughout. To save space, polygons and strips, whose dimensions can be obtained from any of the tessellations, are not shown separately. Thus, a strip of squares can be obtained from Fig. B15. To save more space, a single drawing sometimes contains lines belonging to two strips or tessellations, distinguished by making some of them broken. Thus, the dimensions of any of the Latin-cross strips used can be copied from Fig. B13.

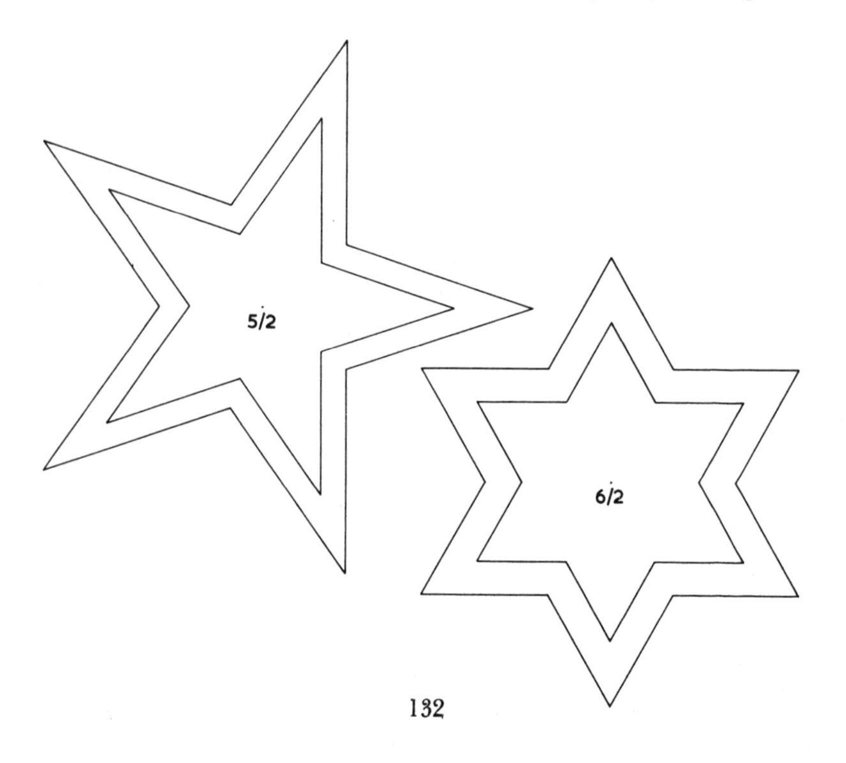

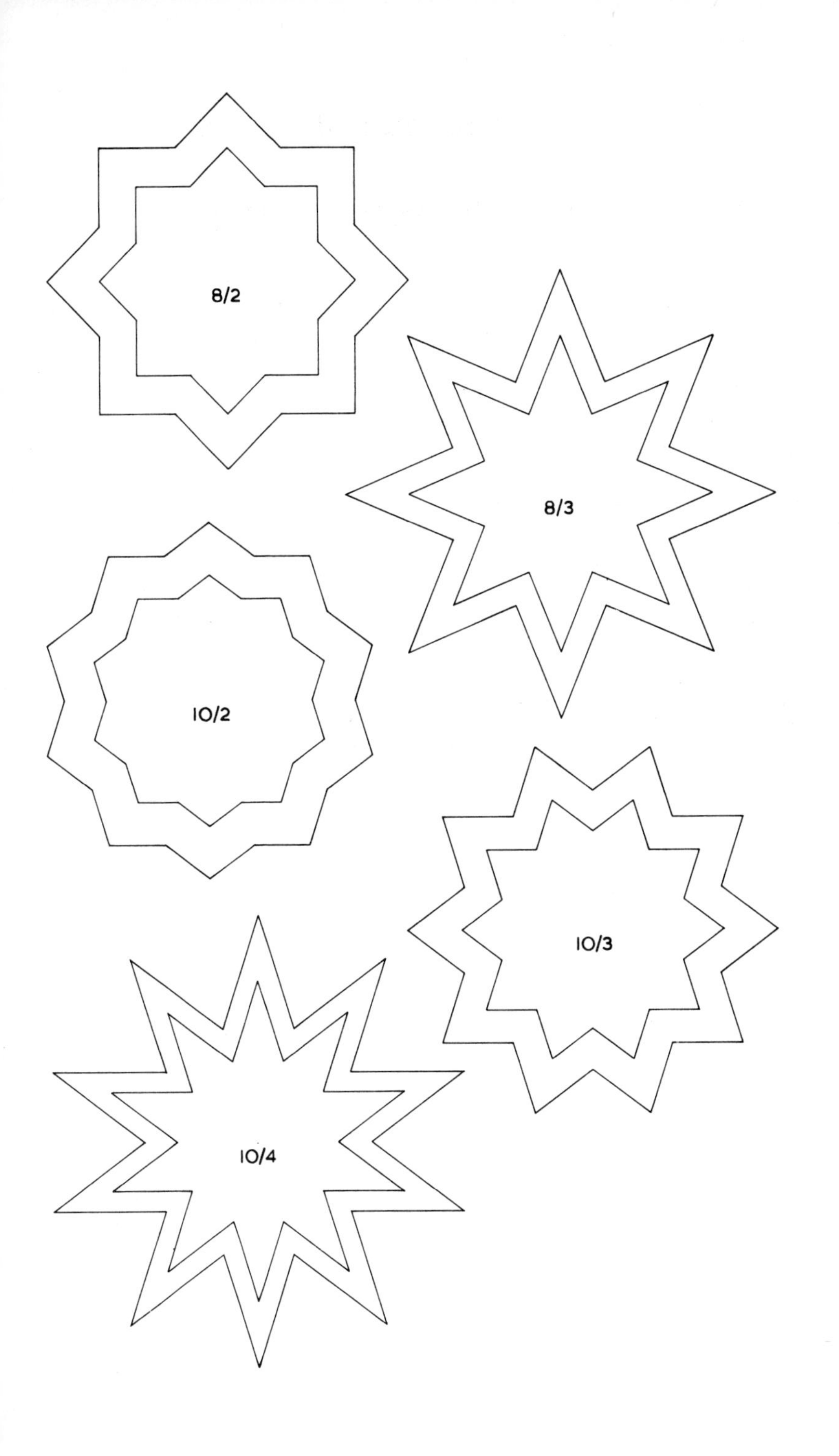
8/2
8/3
10/2
10/3
10/4

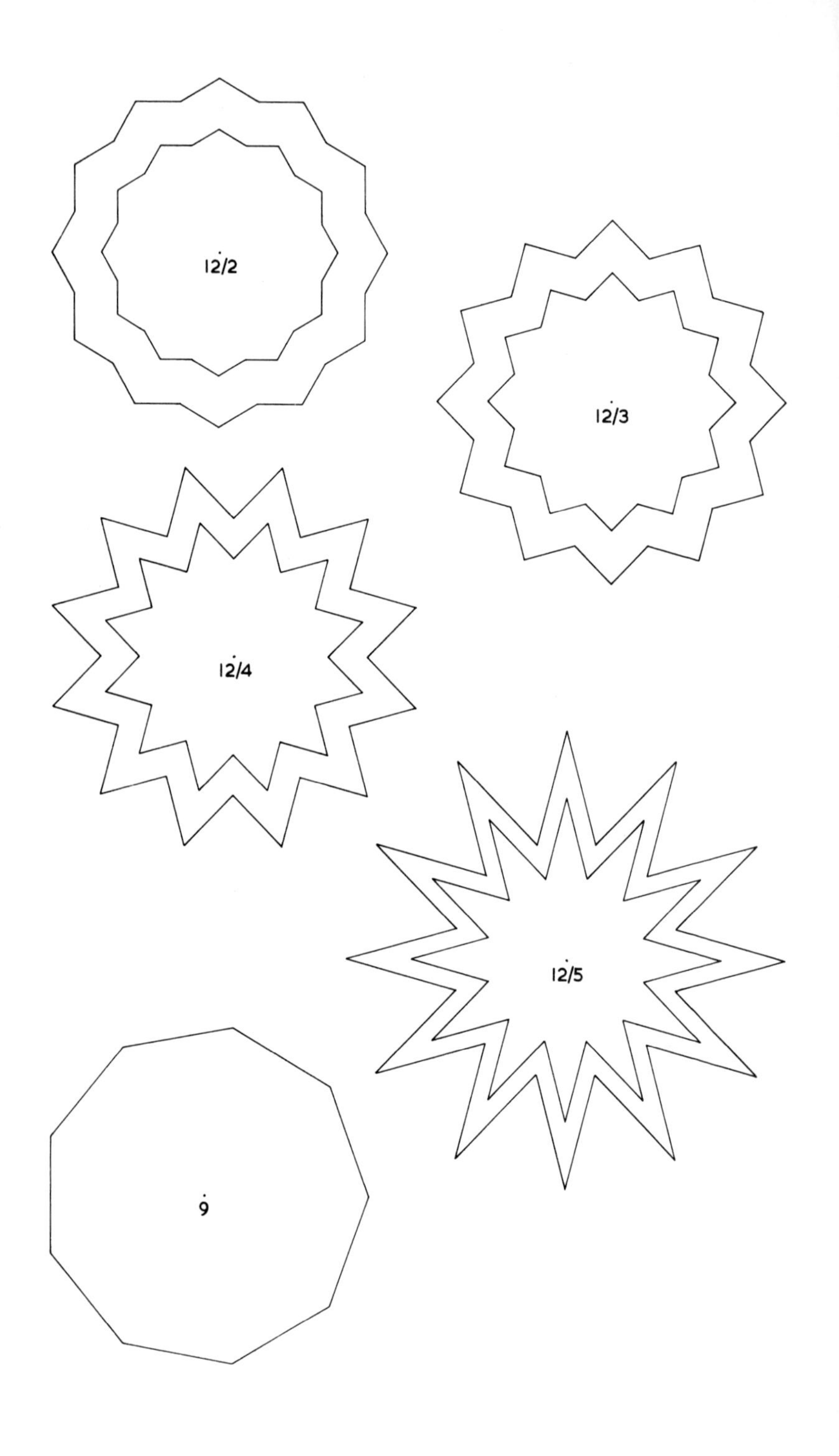

12̇/2
12̇/3
12̇/4
12̇/5
9̇

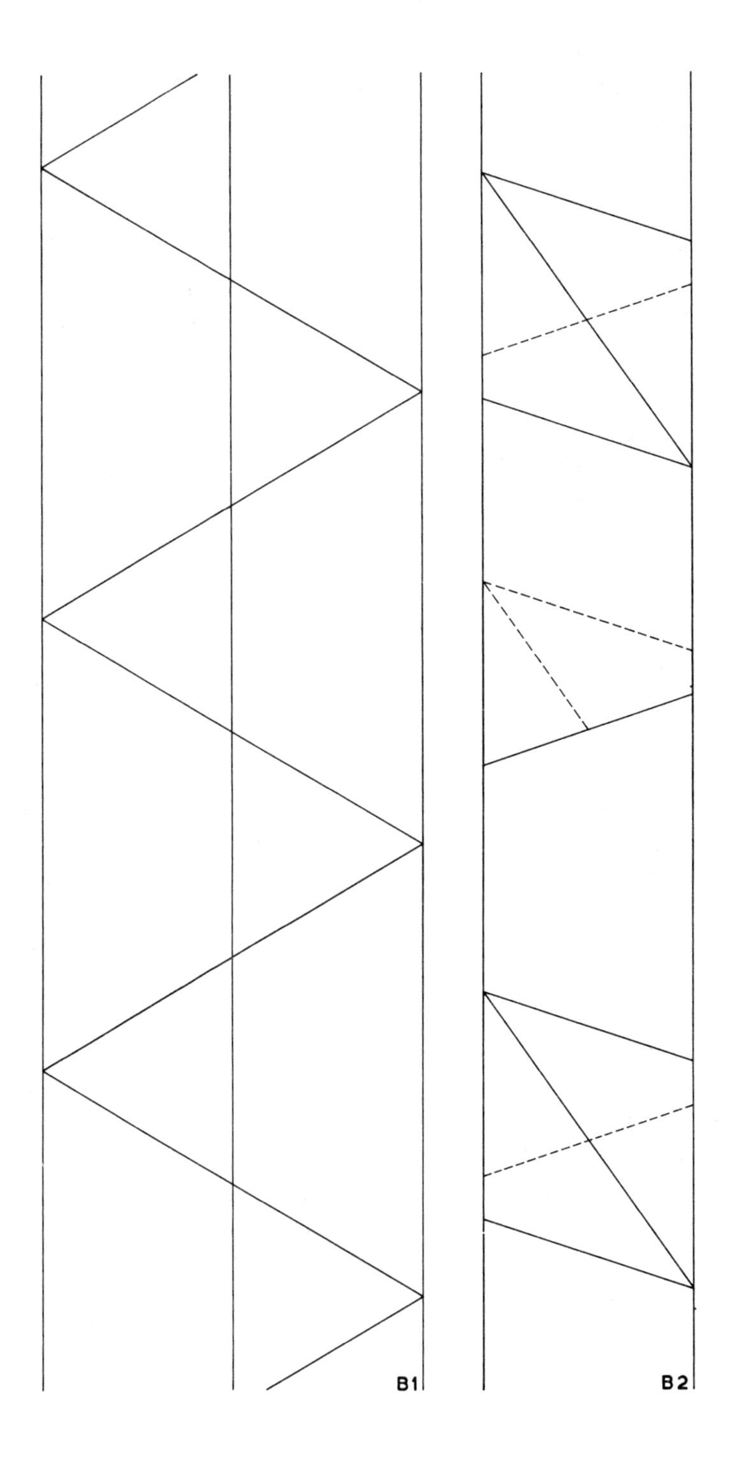
B1
B2

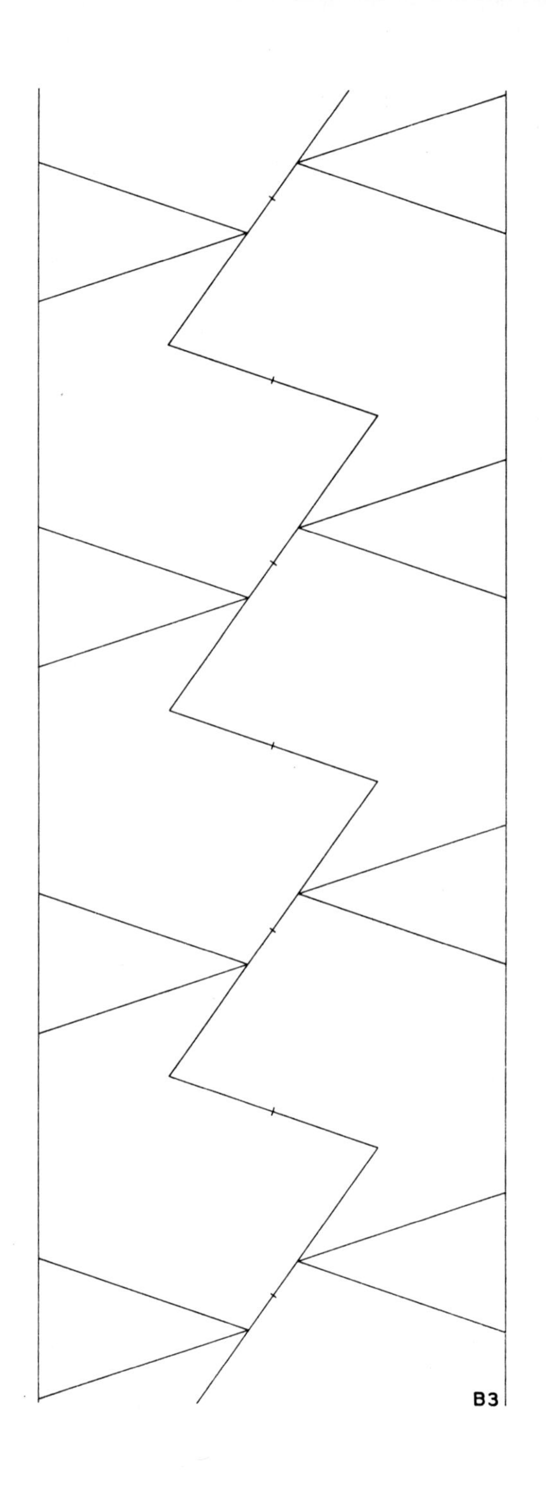
B3

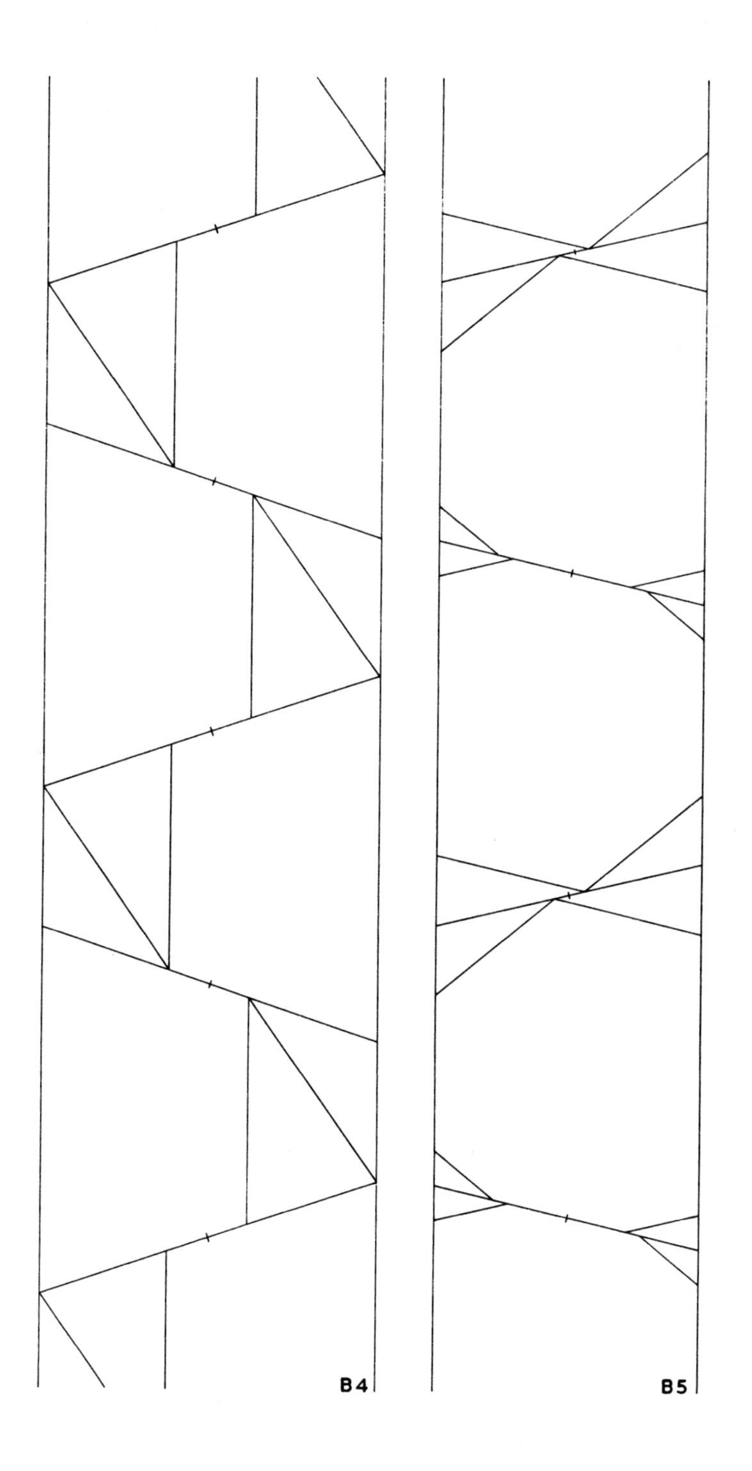
B4
B5

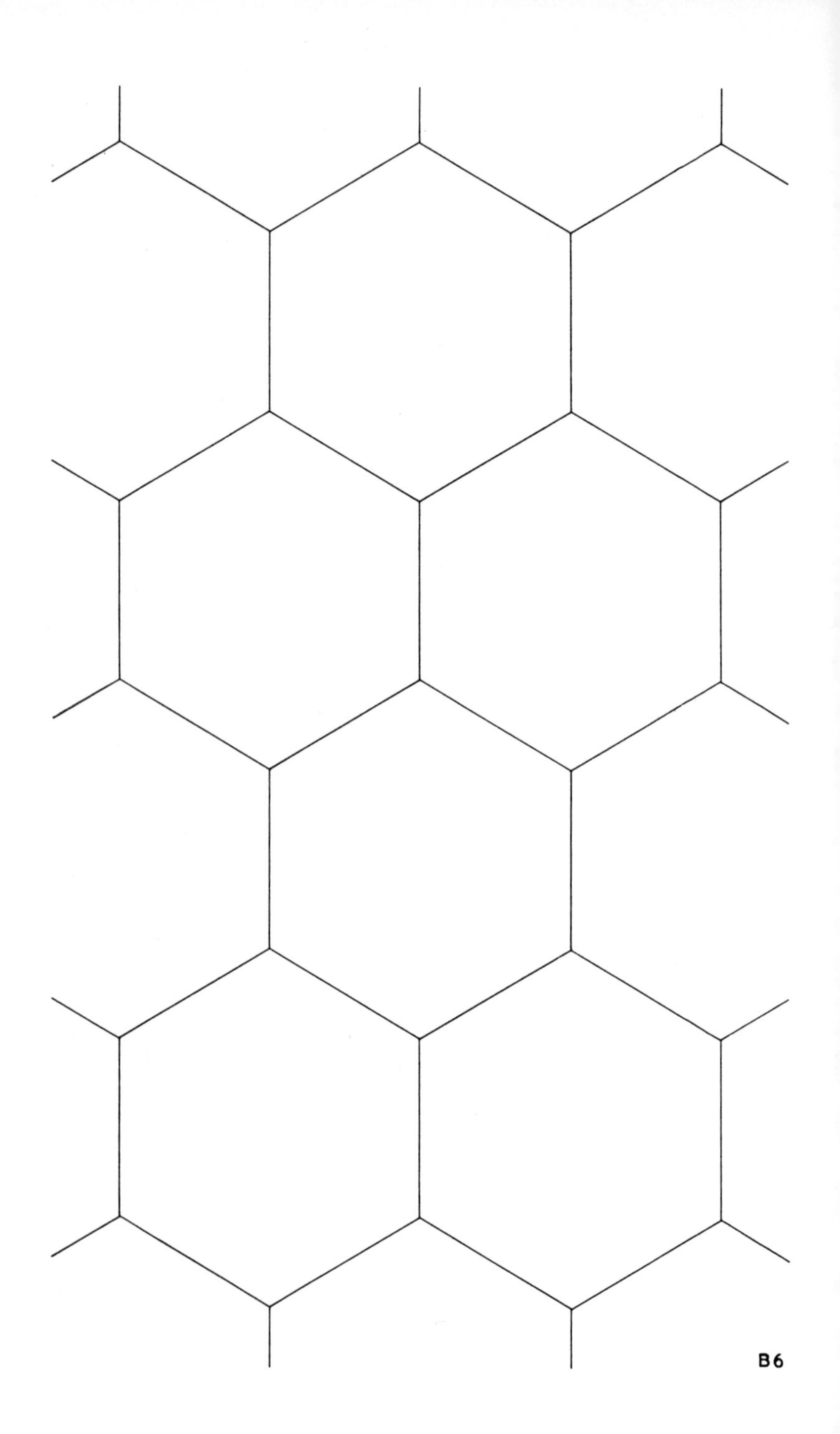
B6

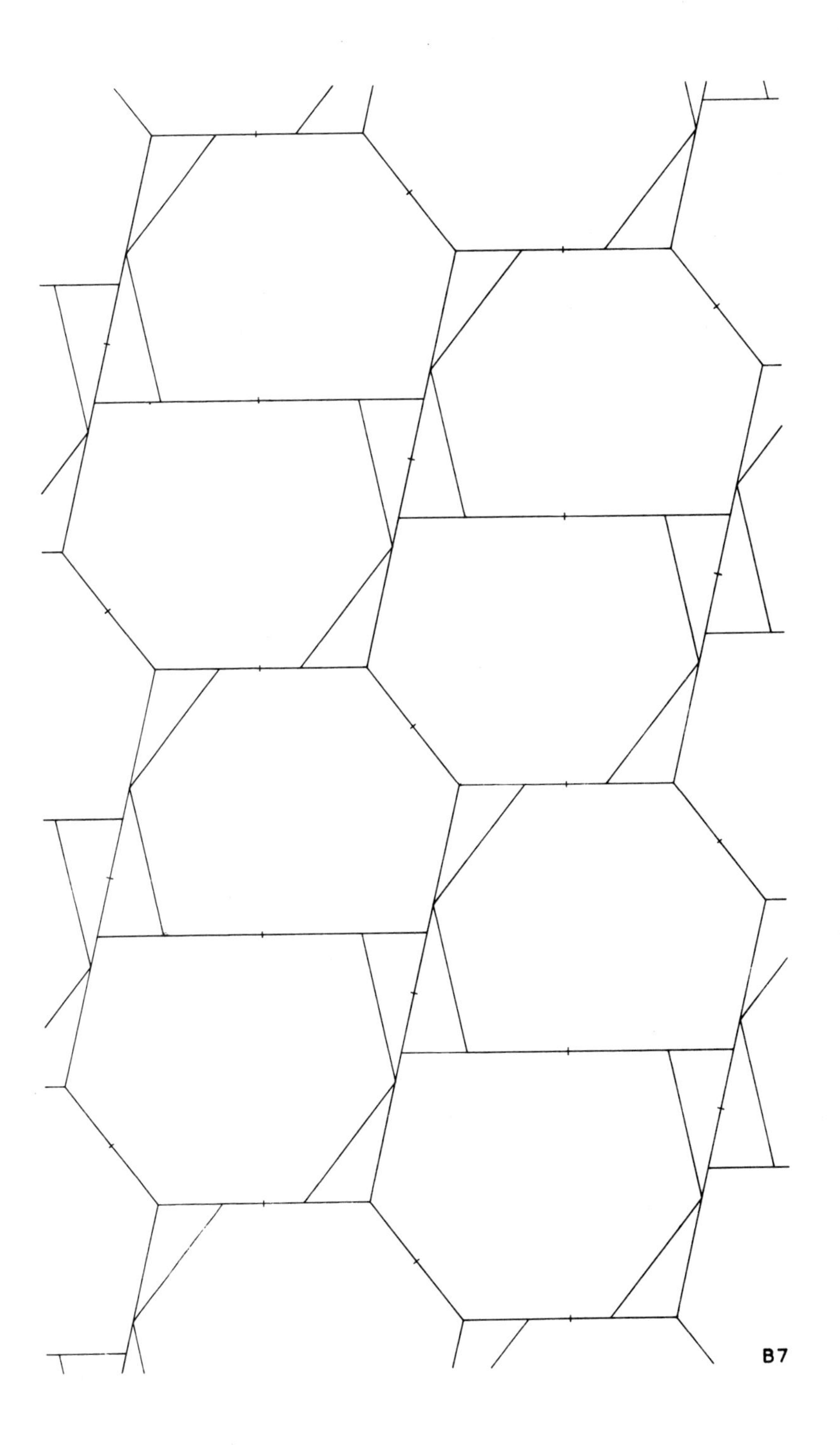
B7

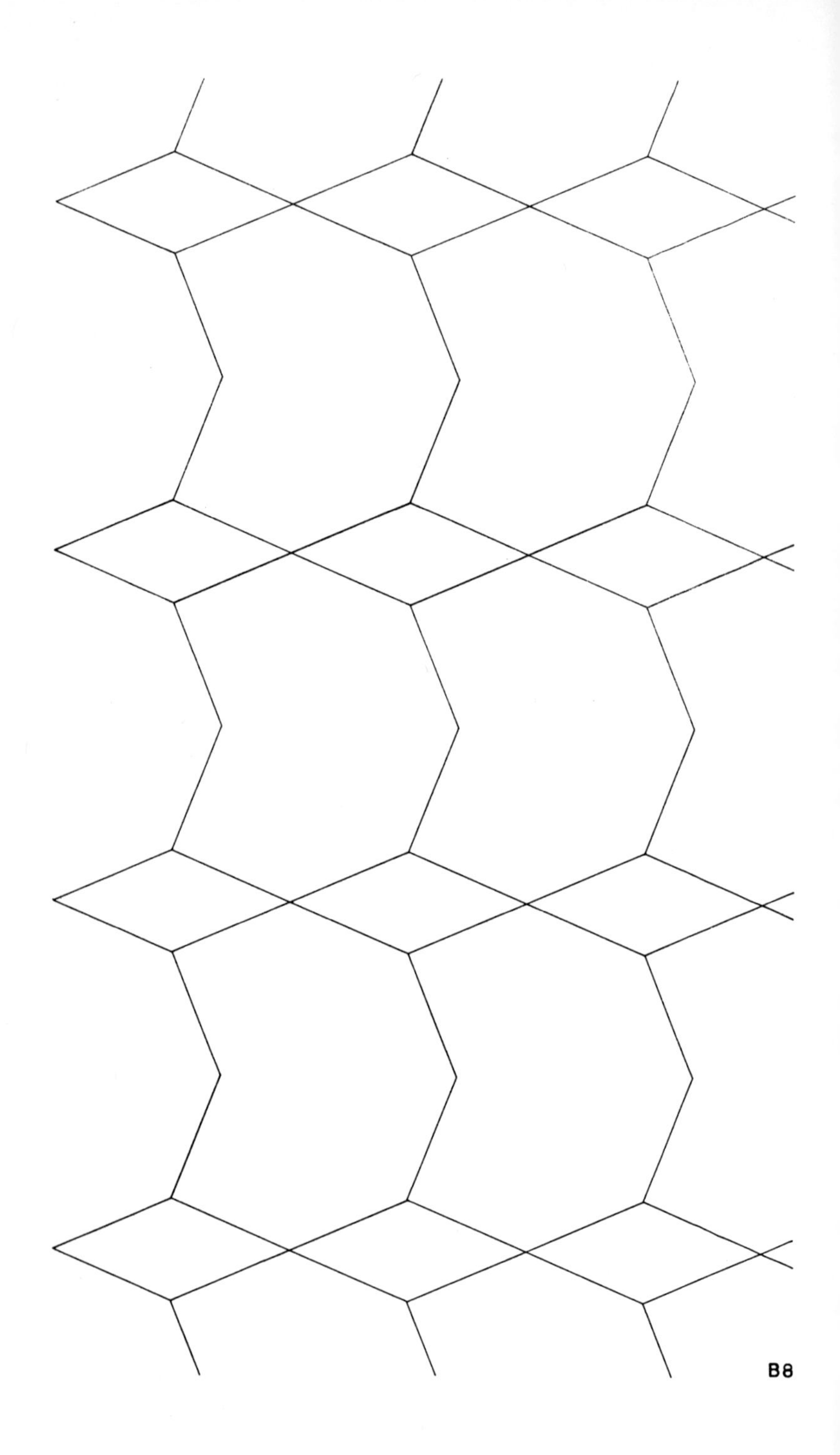

B8

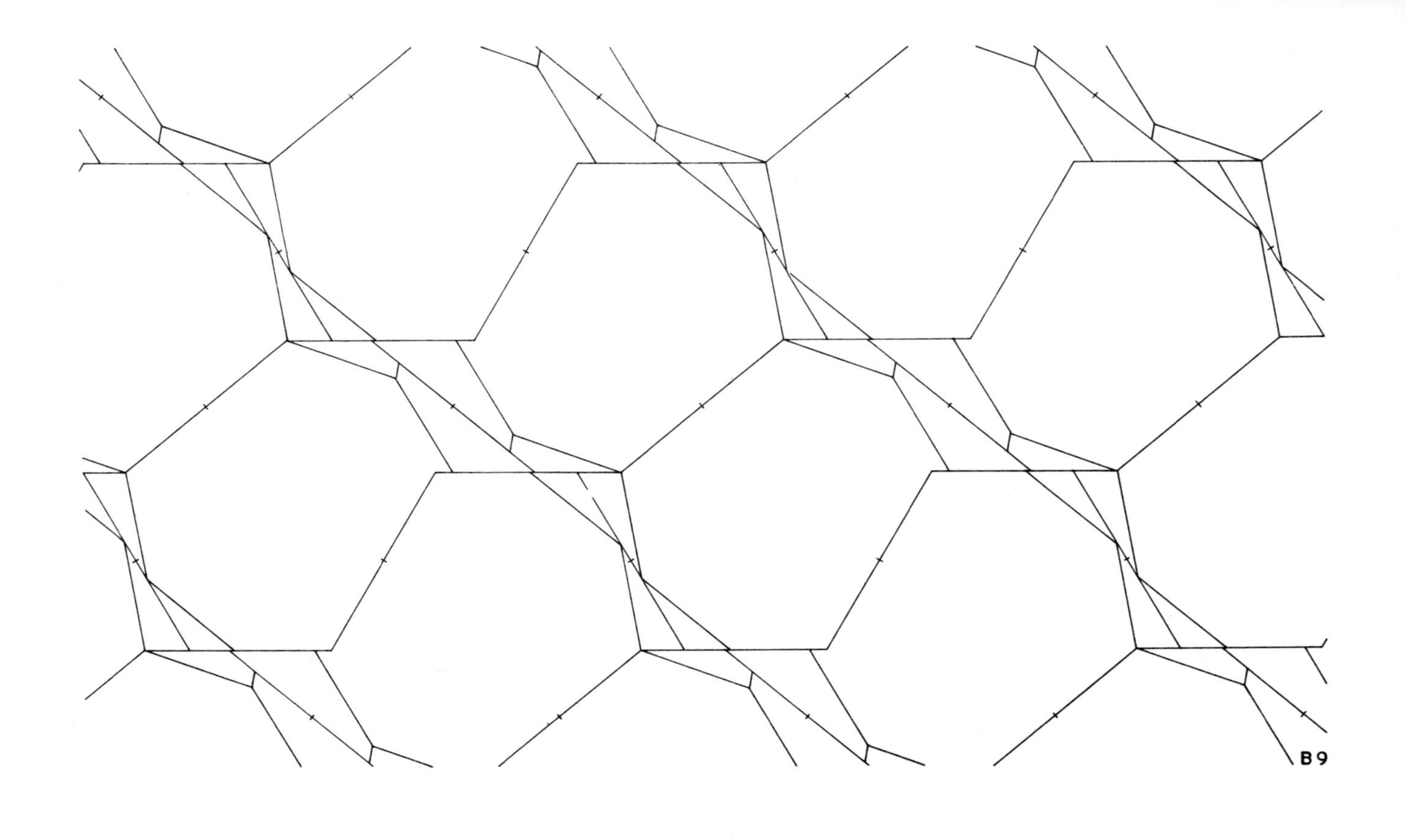
B9

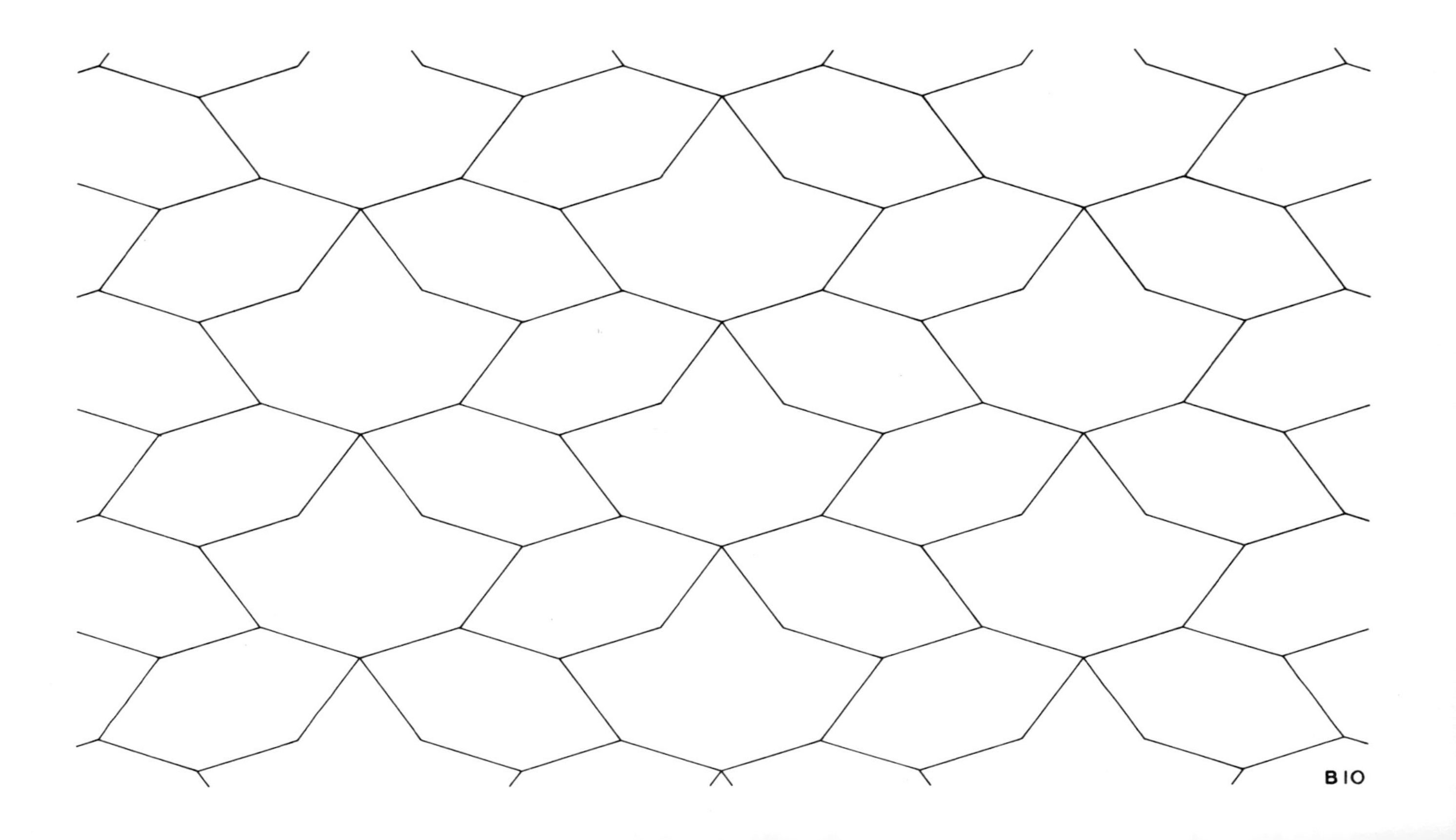

B10

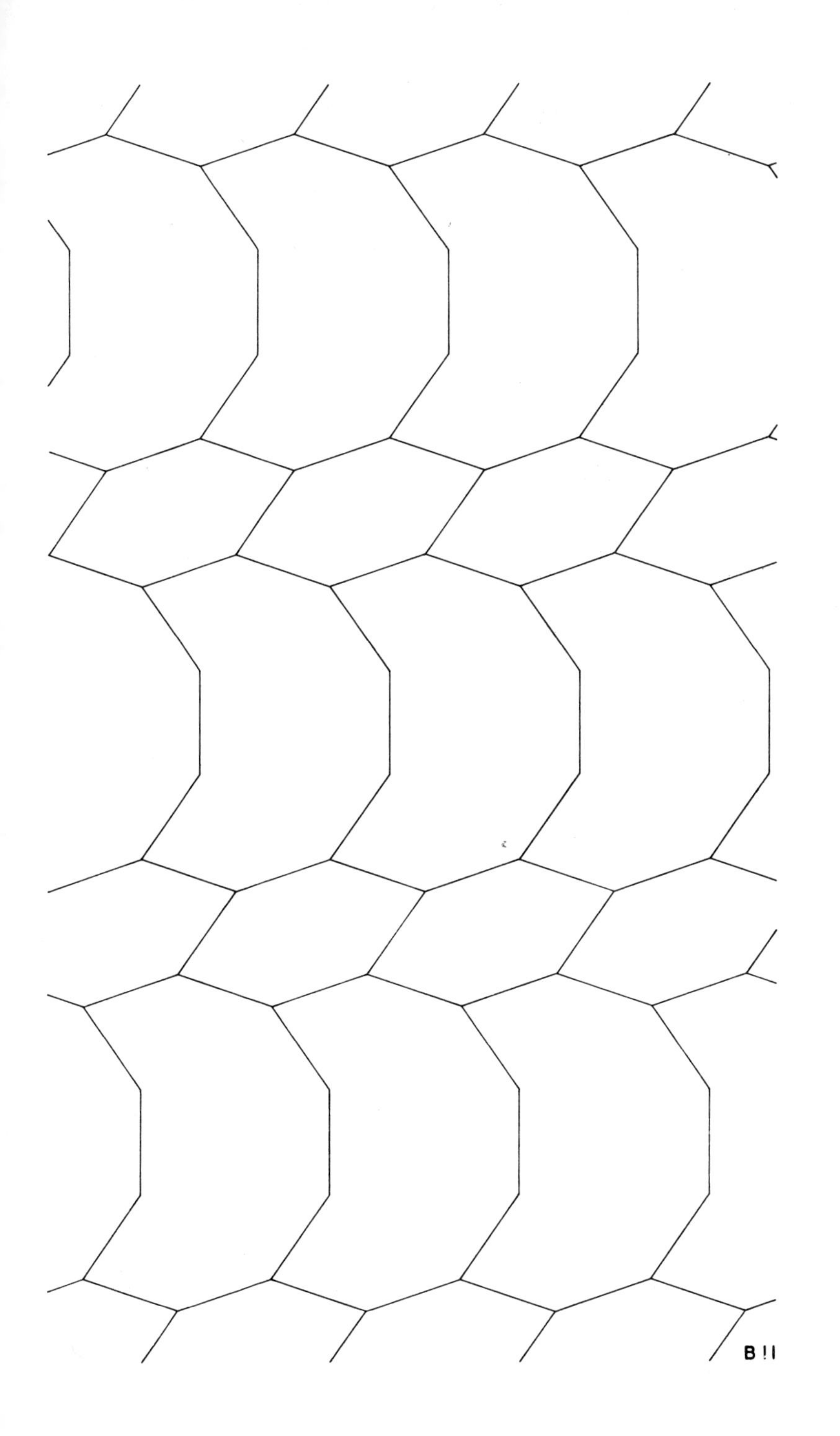

B11

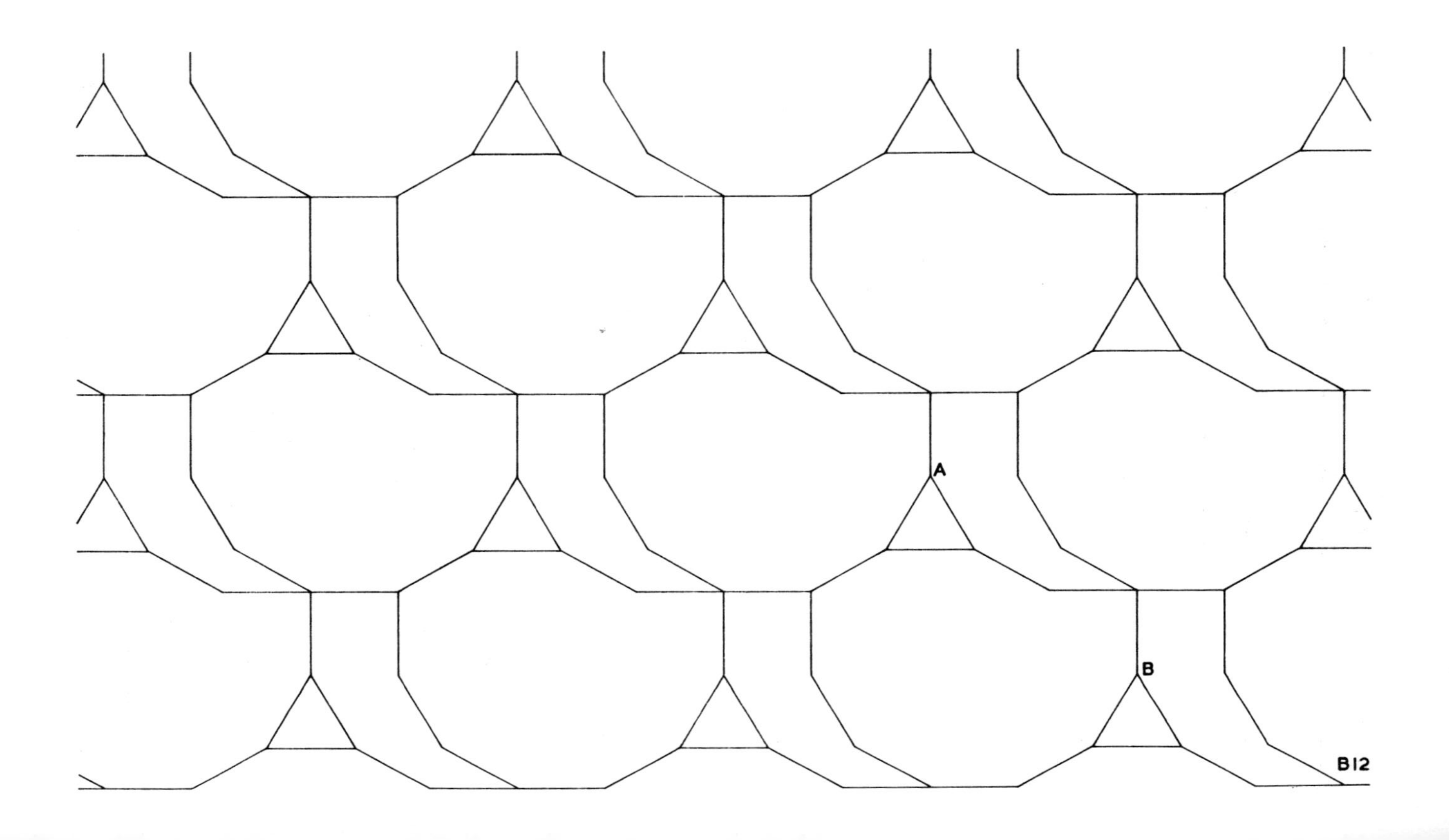
A
B
B12

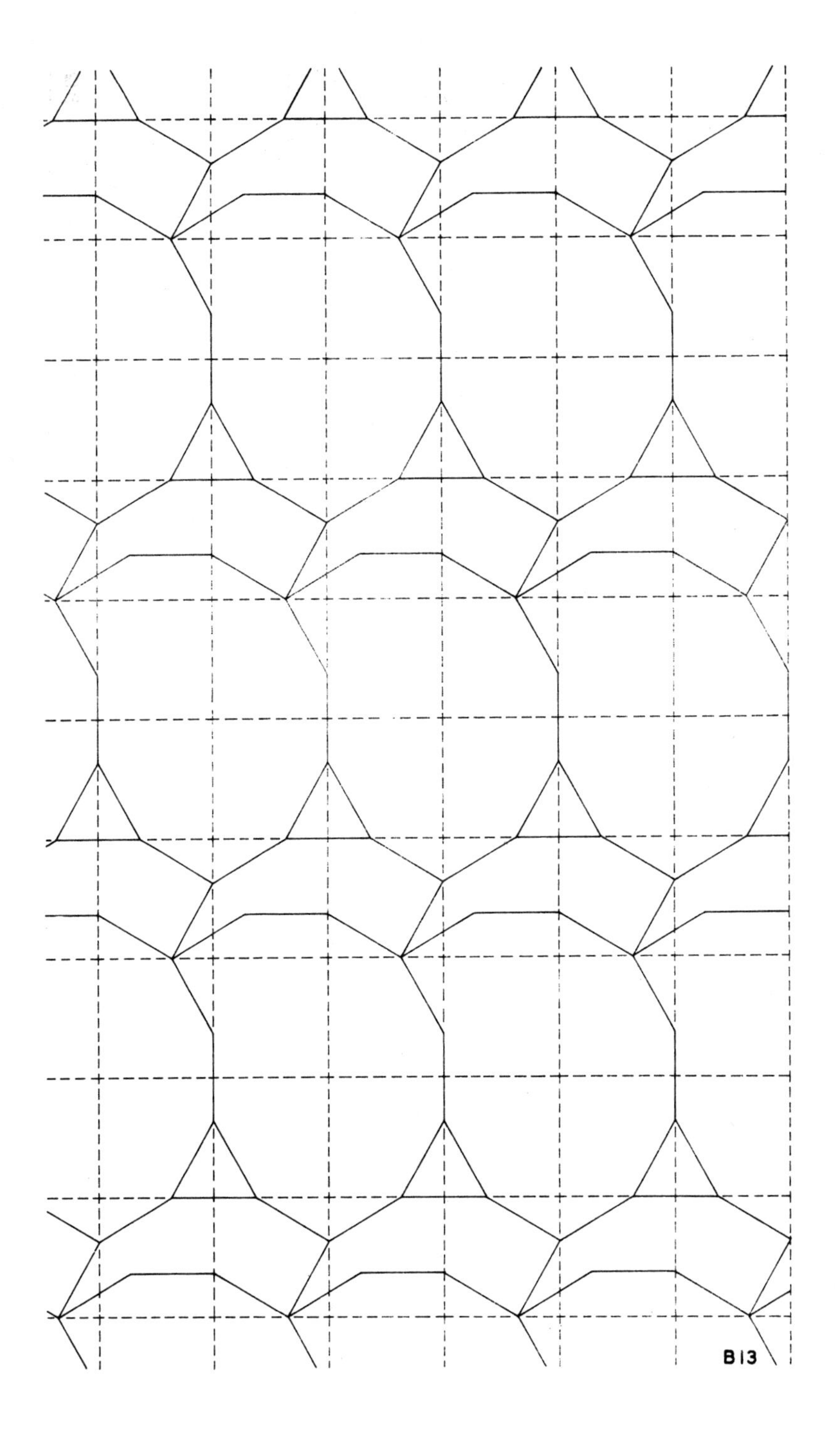
B13

B14

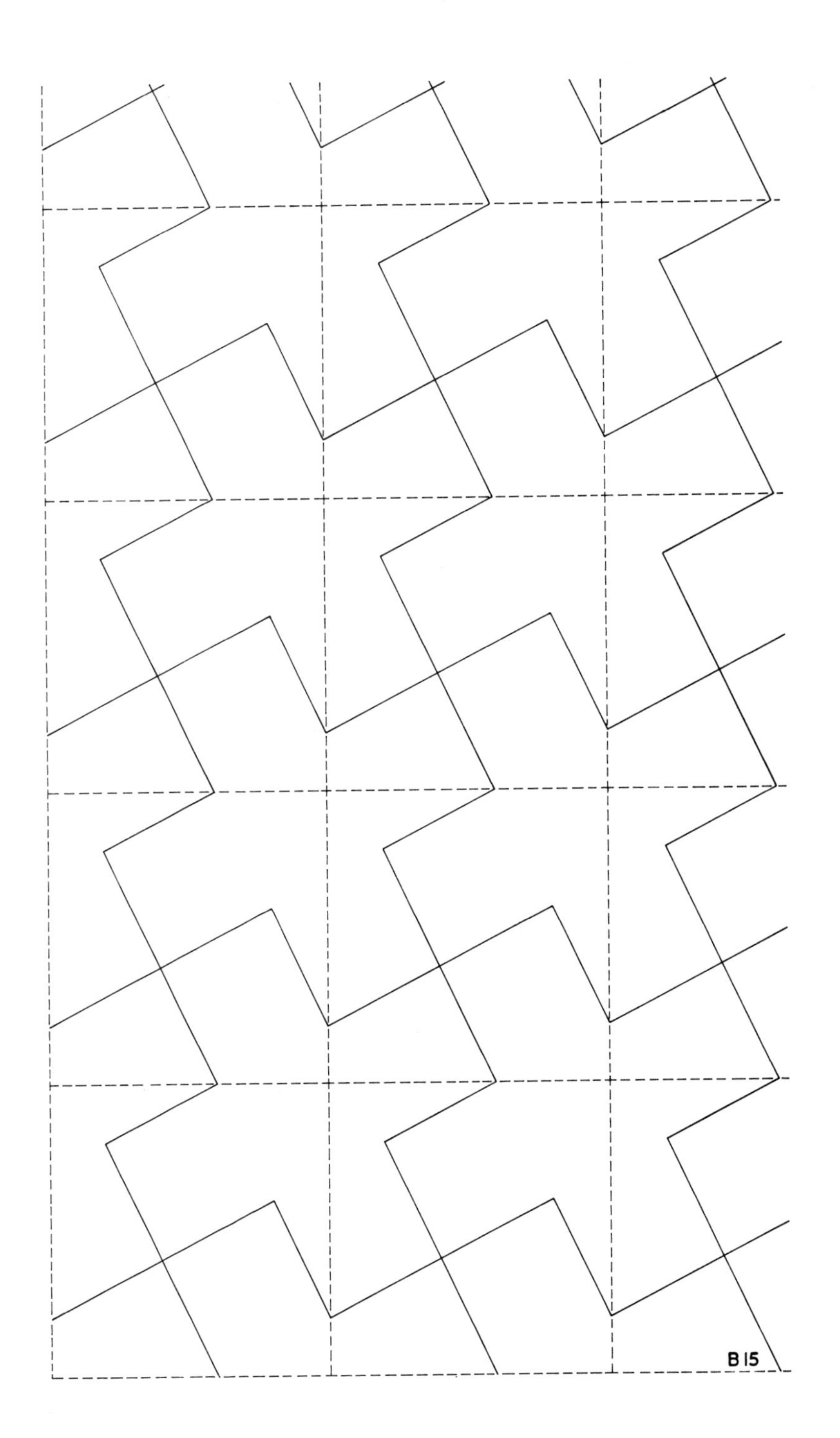
B15

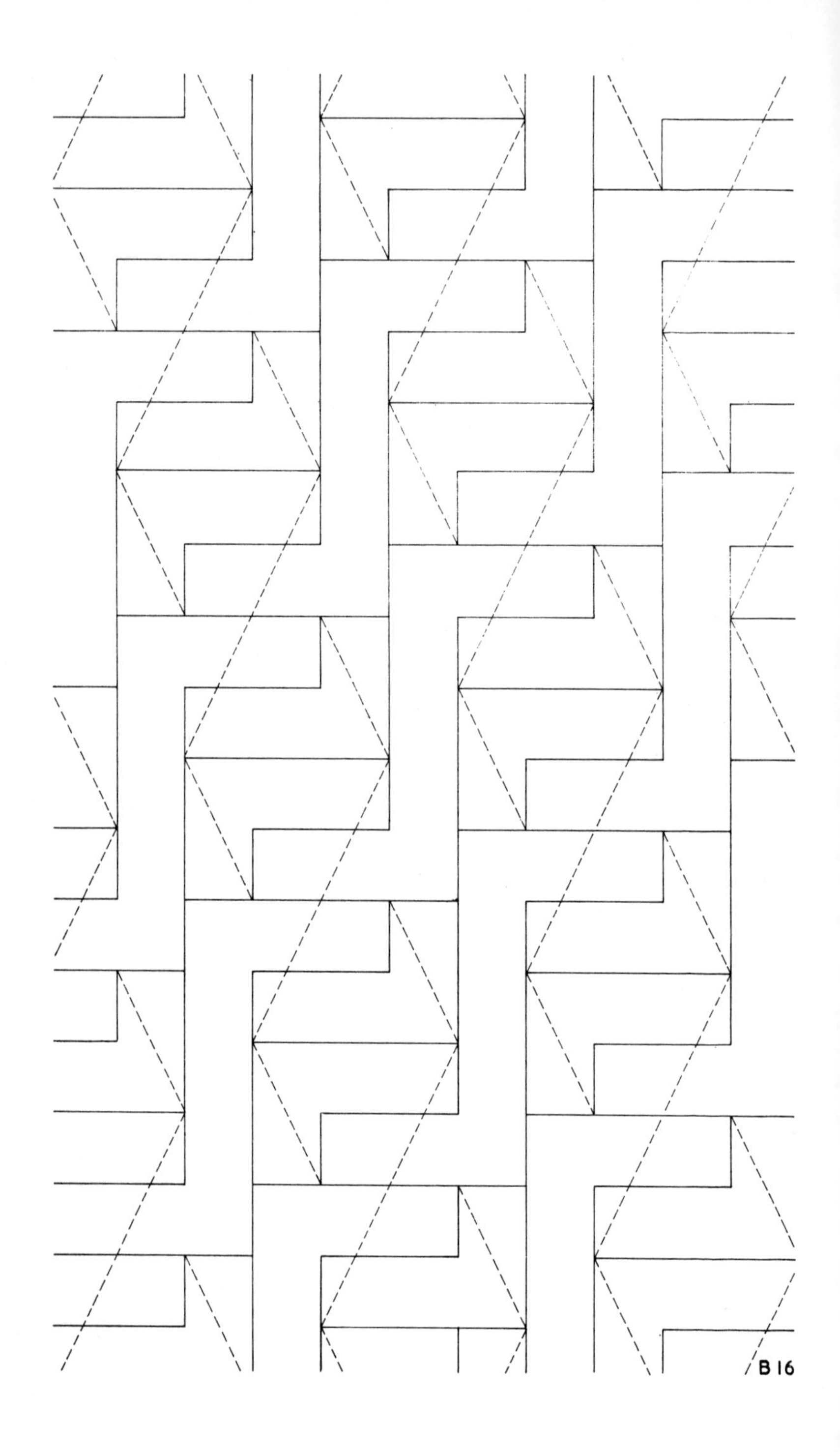
B 16

Appendix C

DIMENSIONS OF POLYGONS

The data in this appendix are for the convenience of those who wish to make accurate drawings. They are for an area of 20 square units. Seven-figure tables were used, to ensure five-figure accuracy.

n	s	log s	R	log R
3	6.7962	83226	3.9238	59370
4	4.4721	65052	3.1623	50000
5	3.4095	53269	2.9003	46244
6	2.7745	44319	2.7745	44319
7	2.3460	37033	2.7035	43192
8	2.0352	30861	2.6591	42474
9	1.7987	25496	2.6295	41988
10	1.6123	20743	2.6087	41642
12	1.3365	12598	2.5820	41195
∞	......	......	2.5231	40194

n	θ	$s \sin \theta$	log	$s \cos \theta$	log
3	30	3.3981	53123	5.8857	76980
5	18	1.0536	02267	3.2426	51090
	36	2.0041	30191	2.7583	44065
6	30	1.3783	14216	2.4028	38072
7	$\pi/7$	1.0179	00770	2.1137	32504
	$2\pi/7$	1.8342	26344	1.4627	16516
	$3\pi/7$	2.2872	35930	0.5220	71770
8	$22\frac{1}{2}$	0.7788	89145	1.8803	27423
	45	1.4391	15810	1.4391	15810

n	θ	$s \sin \theta$	log	$s \cos \theta$	log
9	10	0.3123	49463	1.7714	24831
	20	0.6152	78901	1.6902	22794
	30	0.8993	95393	1.5577	19249
	40	1.1562	06302	1.3779	13921
10	18	0.4982	69742	1.5333	18564
	36	0.9477	97665	1.3043	11539
12	15	0.3459	53898	1.2910	11092
	30	0.6683	82495	1.1575	06351
	45	0.9451	97547	0.9451	97547

n/m	s	log s	R	log R	r	log r
5/2	3.0667	48667	4.2209	62541	1.6123	20743
6/2	1.9619	29267	3.3981	53123	1.9619	29267
7/2	1.4855	17188	3.0847	48922	2.1347	32934
7/3	2.9892	47555	4.2954	63301	1.5330	18555
8/2	1.2102	08284	2.9216	46562	2.2361	34949
8/3	1.8803	27423	3.4743	54087	1.8803	27423
9/2	1.0275	01180	2.8231	45073	2.3015	36200
9/3	1.4087	14881	3.1551	49901	2.0593	31373
9/4	2.9587	47110	4.3253	63602	1.5022	17672
10/2	0.8962	95242	2.7583	44065	2.3464	37040
10/3	1.1400	05692	2.9846	47489	2.1685	33615
10/4	1.8446	26591	3.5087	54514	1.8446	26591
12/2	0.7181	85619	2.6800	42813	2.4028	38072
12/3	0.8393	92391	2.8083	44845	2.2930	36040
12/4	1.1046	04320	3.0178	47968	2.1339	32917
12/5	1.8257	26144	3.5271	54741	1.8257	26144

The sides of the unit square for the various crosses are:

	s	log s
G	2	30103
L	1.8257	26144
M, S	1.0847	03529

The dimensions of a golden rectangle are 5.6886 × 3.5158 (log = 75501 + 54602).

Appendix D

SOLUTIONS OF PROBLEMS

The solutions given here of the problems in Appendix A include only the common area of *S*-dissections and the rearrangements of *P*-slides are omitted. Why? See the end of Chapter 4.

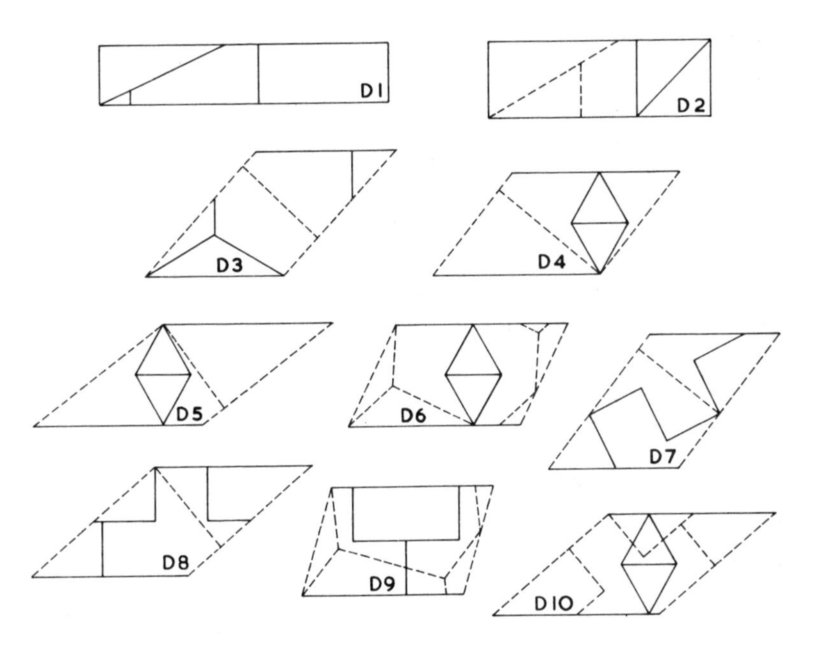

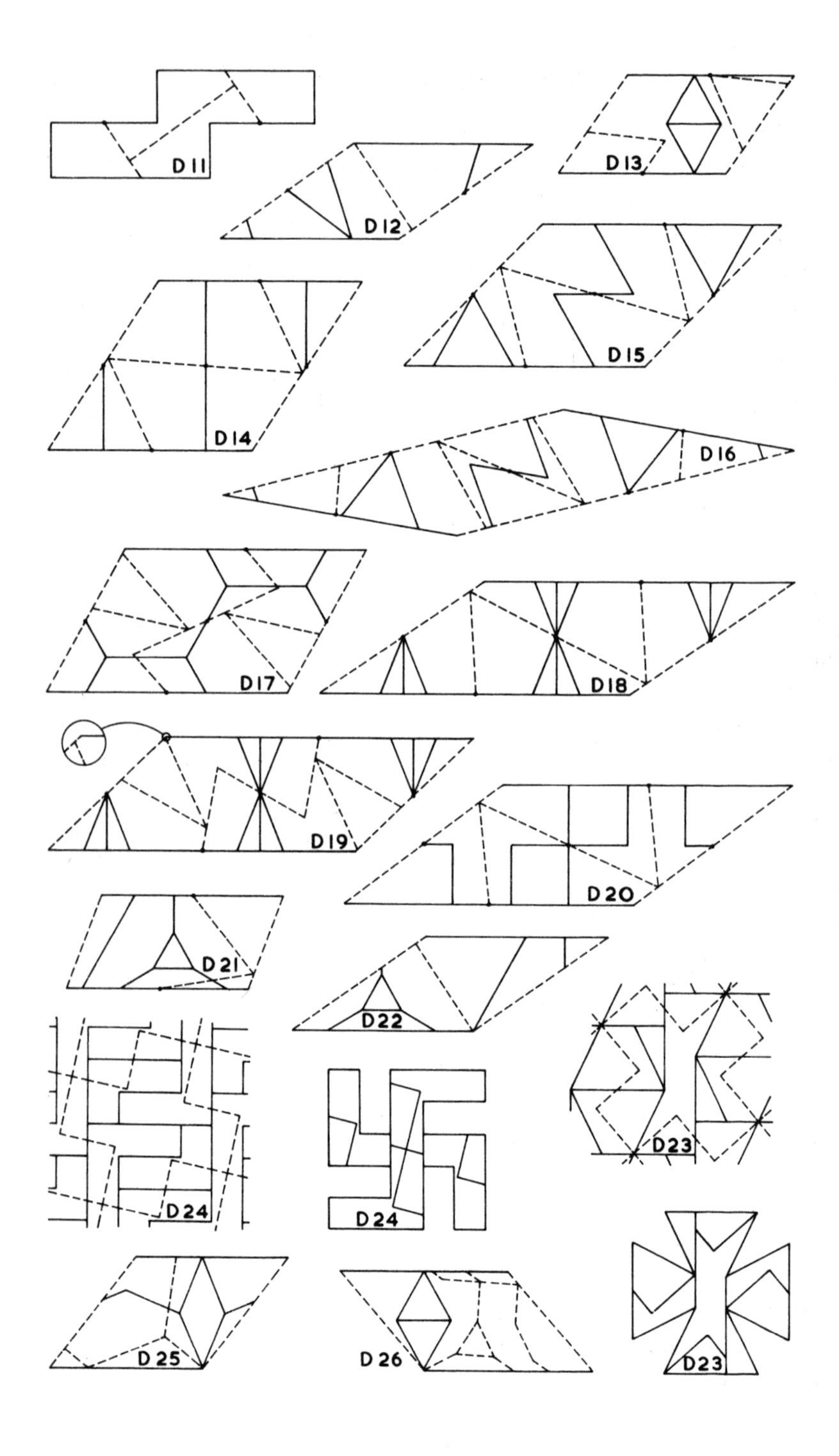
D11
D12
D13
D15
D14
D16
D17
D18
D19
D20
D21
D22
D23
D24
D24
D25
D26
D23

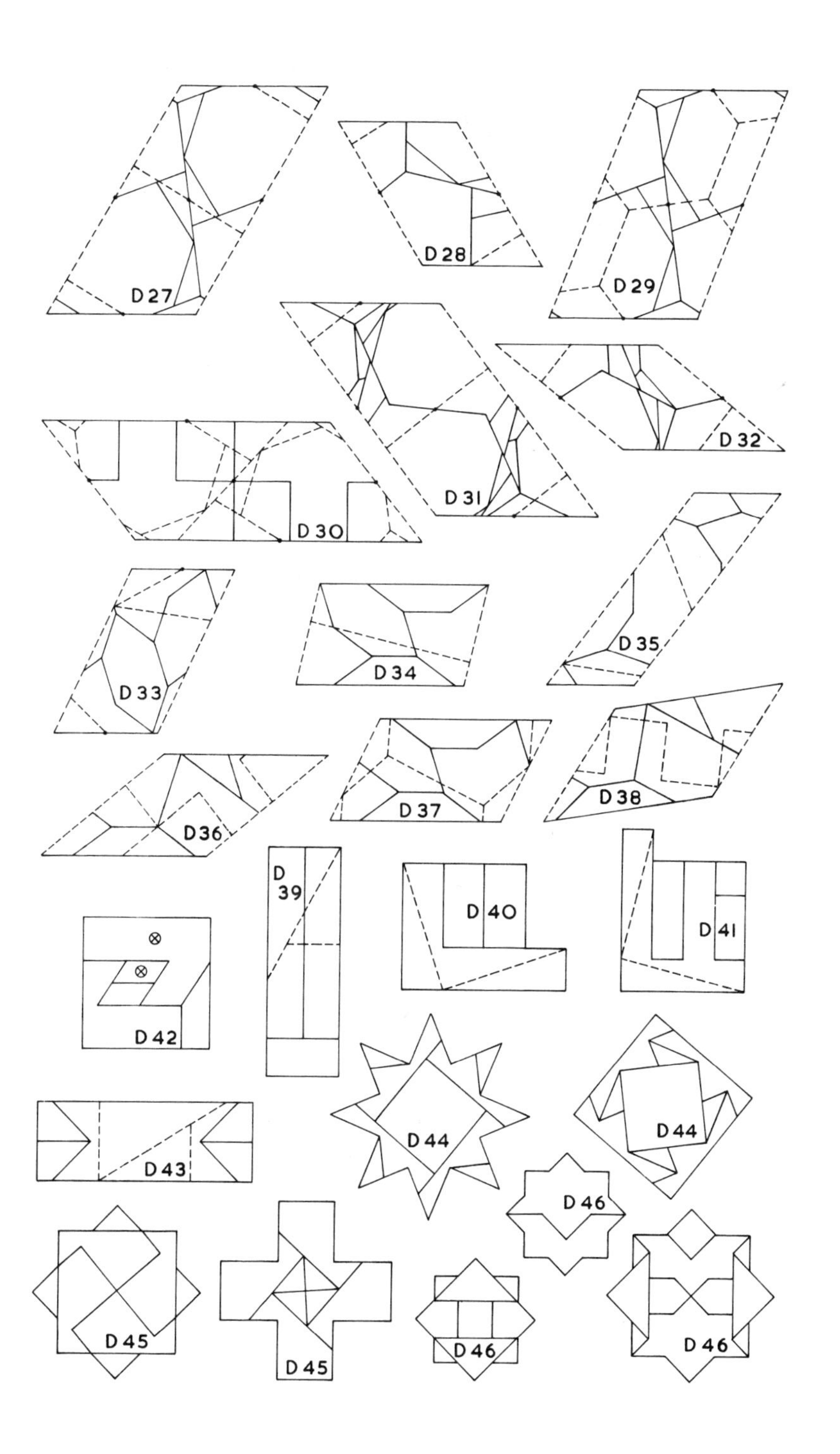
D 27
D 28
D 29
D 31
D 32
D 30
D 33
D 34
D 35
D 36
D 37
D 38
D 39
D 40
D 41
D 42
D 43
D 44
D 44
D 46
D 45
D 45
D 46
D 46

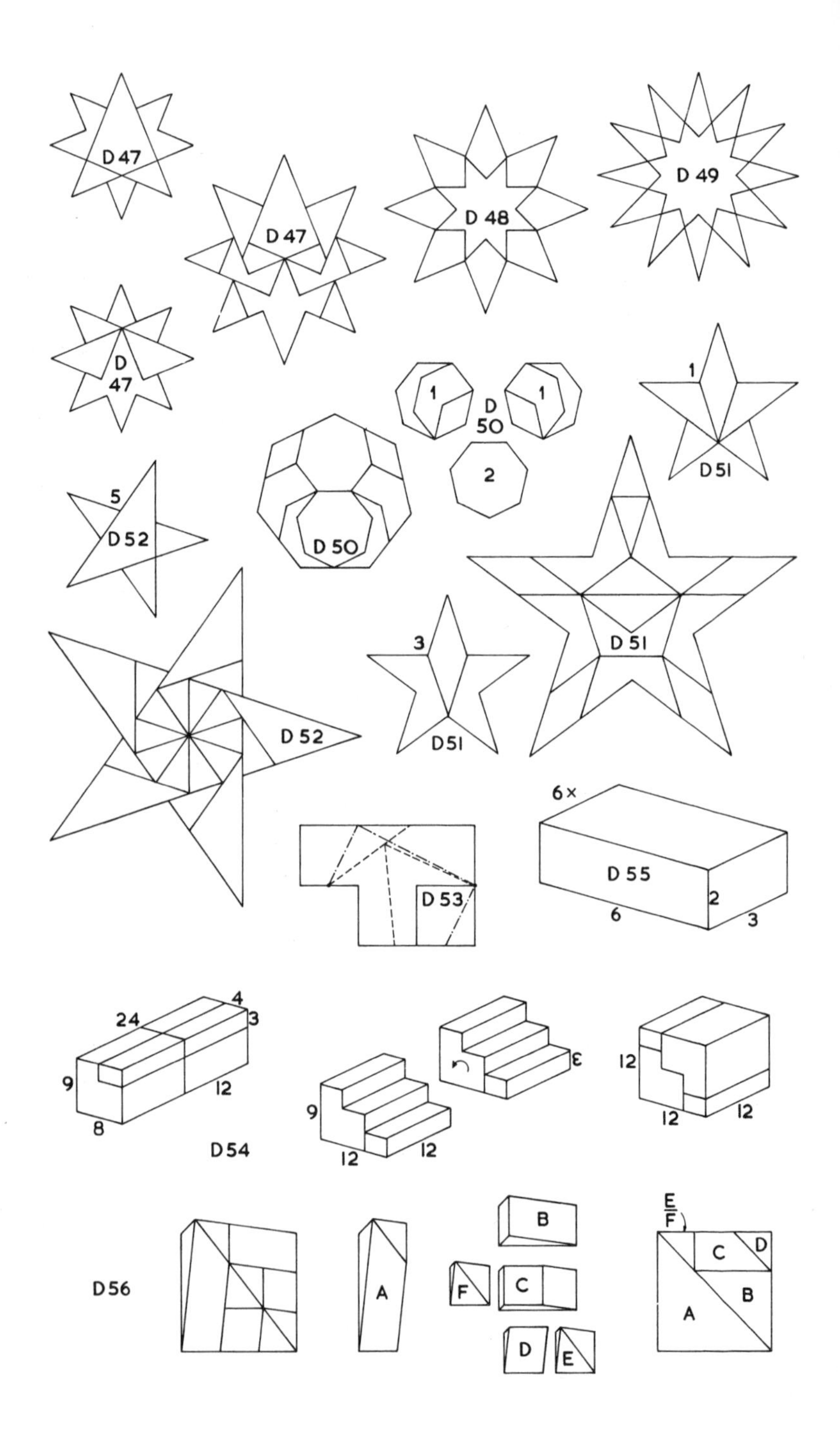

D 47
D 47
D 48
D 49
D 47
1
1
D 50
1
D 51
2
5
D 52
D 50
3
D 51
D 52
D51
6×
D 53
D 55
2
6
3
4
24
3
9
12
8
D54
9
12
12
3
12
12
12
D56
B
E
F
C
D
F
C
A
B
A
D
E

Appendix E

LIST OF DISSECTIONS

The following list comprises all dissections involving only regular polygons, polygrams, and crosses, and so will serve as an index to the vast majority of the dissections described. Dissections of letters, with one exception, are all in Chapter 18 and Figs. D39-D42. To list those that include other irregular polygons was found impractical.

The numbers and letters in columns A and B denote the polygons dissected. For appearance' sake a regular *n*-gon is denoted by *n* instead of $\{n\}$, and a regular $\{n/m\}$ by n/m. **R** denotes a golden rectangle, G, L, M denote Greek, Latin, Maltese cross, and S_1 and S_2 denote swastikas. Two equal *n*-gons are denoted by n^2, two unequal ones by $n \cdot n$, and so on.

The number of pieces is in all cases minimal. Thus, the third column indicates the present record, and constitutes a challenge to the reader. In none of the dissections listed does any piece need to be turned over.

If the first-named polygon in a *PT* dissection has the *T*-strip, this is indicated in the column headed Type by *TP*.

A	B	Pieces	Type	Figure
3^3	3	6	*T*	21.13
3^4	3	4	*T*	21.4
3^7	3	13	*T*	21.14
$3 \cdot 3$	3	5	*P*	1.9
			Q	1.11
4	3	4	*TT*2	5.2
4^2	4	4	*T*	21.23
4^3	4	6	*P*	1.6
$4 \cdot 4$	4	5	*T*	9.17
R	3	4	*TT*2	D14
R · R	R	5	*T*	21.30
5	3	6	*PT*	4.8
			*TT*1	5.1

A	B	Pieces	Type	Figure
5	4	6	*PP*	3.1
			TP	4.2
			*TT*2	5.6
	R	6	*TP*	D12
5^2	5	11	*U*	21.27
5^4	5	6	*U*	21.4
5^5	5	12	*U*	21.24
5 • 5	5	11	*U*	21.28
6	3	5	*TT*2	D15
	3^2	4	*T*	20.18
	4	5	*PP*	3.2, 3.3, 3.4
	R	5	*PP*	D3
	5	7	*PP*	3.5
			*TT*2	D16, D17
6^2	3	6	*T*	20.19
6^2	6	9	*T*	21.34
6^3	6	6	*T*	11.2
6^7	6	12	*T*	11.3
6^7	6^3	18	*T*	11.4
6 • 6	6	8, 9	*T*	21.32, 21.34
	6	5	*R*	H32
7	3	9	*TT*2	15.1
	4	9	*PP* + *TT*2	6.4
			*TT*2	D27
	R	9	*TP*	D28
	5	11	*TT*2	15.8
	6	11	*TP*	15.9
			*TT*2	D29
7^4	7	8	*U*	D50
8	3	8	*TT*2	D18
	4	5	*T*	12.1
	R	6	*PP*	2.23
	5	9	*PP*	3.6
			*TT*2	D19
	6	9	*PP*	2.11, D25
	7	13	*PP*	15.7
8^2	8	8	*U*	21.16, 21.17
8^8	8	24	*U*	21.18
9	3	9	*PT*	16.10 - 16.13
	4	12	*TP*	16.8
			*TT*2	D31

A	B	Pieces	Type	Figure
9	R	12	*TP*	D32
	6	14	*TT*2	16.9
9^3	9	15	*U*	H18
9^4	9	10	*U*	21.5
10	3	8	*PP*	G5
	4	8	*PP*	17.2, 17.6, 17.9
	R	7	*PP*	G6
	5	10	*PT*	G4
	6	9	*PP*	D35
	7	13	*PT*	H1
	8	12	*PP*	G7, G8
10^5	10	18	*U*	H20
12	3	8	*PT*	D21
	4	6	*T*	9.7, 9.8, 9.9, 9.10
	R	7	*PP*	D22
	6	6	*T*	12.2
12^2	12	10	*U*	21.15
12^3	12	15	*U*	21.21
12^3	12^2	23	*U*	21.26
12^4	12	12	*U*	21.6
12^6	12	24	*U*	H21
15^3	15	24	*U*	H19
16^2	16	16	*U*	H17
5/2	3	9	*TT*2	19.3, see text
	4	8	*PP*	6.5
	6	10	*TT*2	19.2
	10	6	*U*	20.17
$5/2^4$	5/2	14	*U*	D51
$5/2^5$	5/2	20	*U*	D52
6/2	3	5	*T*	19.4, 19.5
	4	5	*PP*	D4
	R	5	*PP*	D5
	5	8	*PT*	4.9
	6	7	*PP*	3.9
			PT	D13
	6^2	4	*T*	20.7
	7	11	*PT*	15.10
	8	9	*PP*	D6
	10	9	*PP*	G9
	12	10	*PP*	D26
$6/2^2$	6	6	*T*	20.8

A	B	Pieces	Type	Figure
$6/2^3$	6/2	12	*U*	H 22,23
$6/2 \cdot 6/2$	6/2	15	*T*	21.31
7/3	7^2	9	*U*	H12
$7/3^2$	7	14	*U*	H13
8/2	4	7	*T*	H2
$8/2^2$	8/2	11	*U*	D46
8/3	4	8	*P*	D43
	8	6	*U*	20.5
	8^2	9	*U*	20.9
$8/3^2$	8	13	*U*	20.10
	8/3	12	*U*	D47
$8/3 \cdot 8/2$	8/3	9	*U*	D48
9/3	6	9	*T*	H14
9/4	9/2	14	*U*	H15
10/2	5	8	*U*	20.3
10/3	10^2	6	*U*	20.13
	$5/2^2$	10	*U*	20.20
$10/3^2$	10	10	*U*	20.14, 20.15
10/4	10^2	11	*U*	20.16
	$5/2^2$	18	*U*	20.21
	10/3	11	*U*	20.16
12/2	3	6	*U*	G14
	6	8	*T*	H3
	6/2	9	*T*	G11
			U	G12, G13
$12/2^2$	12/2	13	*U*	H25
$12/2^3$	12/2	18	*U*	H27
12/3	4	10	*T*	H7
$12/3^2$	12/3	12	*U*	H24
$12/3^3$	12/3	21	*U*	H28
$12/4^2$	12/4	18	*U*	H26
$12/4^3$	12/4	24	*U*	H29
$12/4 \cdot 12/3$	12/5	13	*U*	D49
12/5	4	9	*T*	H5
	12	10	*U*	H4
14/2	7	14	*U*	H9
14/2	$7/3^2$	18	*U*	H11
14/3	$7/2^2$	18	*U*	H8
14/5	14/2	14	*U*	H10
18/7	$9/2^2$	18	*U*	H16
G	3	5	*TT*2	D20

A	B	Pieces	Type	Figure
G	4	4	*T*	9.1
	4 • 4	5	*T*	12.3, 12.4
	R	5	*PP*	D7
	5	7	*TT2*	5.8
	6	7	*PP*	3.7, 22.6
			TP	13.2
	7	12	*TT2*	D30
	8	9	*T*	See Ch. 22
	10	10	*PP*	G10
	12	6	*T*	9.11
	6/2	8	*PP*	3.10
	8/2	7	*T*	H6
G • 4	G	5	*U*	10.1 and 10.3
G^2	G	5	*T*	21.10
G^3	4	9	*P*	1.8
			PP	2.1
	G	12	*PP*	13.4
G^5	G	12	*T*	21.11
	G^2	12	*T*	21.12
L	3	5	*TT2*	5.3
	4	5	*PP*	2.7
	R	5	*PP*	D8
	5	8	*PT*	4.10
	6	6	*PP*	3.8
	7	12	*PT*	15.11
	8	8	*PP*	D9
	10	10	*PP*	D36
	12	7	*T*	9.6
	6/2	9	*PP*	D10
	12/5	7	*U*	20.24
	G	7	*PT*	4.11
M	4	7	*T*	11.5
	6	14	*PP*	13.7
	G	9	*T*	D23
S_1	4	6	*T*	9.14
	6	12	*PP*	13.6
	G	8	*T*	D24
	M	9	*T*	9.12 + 9.13
S_2	4	4	*T*	9.15
	G	4	*T*	9.16

Appendix F

SOURCES AND CREDITS

The works in which most of the "prior art" was found were Henry E. Dudeney, *Amusements in Mathematics; The Canterbury Puzzles; Modern Puzzles; Puzzles and Curious Problems; A Puzzle-mine;* and Martin Gardner, *Mathematical Puzzles of Sam Loyd,* Vols. I and II. These are referred to below by the initials *AM, CP, MP, PCP, PM, SL*1, *SL*2, followed by the puzzle number; *AM* Fig. x denotes Fig. x in the article "Greek Cross Puzzles," in *AM*.

A search was also made in *The Mathematical Gazette* (referred to as *MG*) back to the beginning, and the problems in *The American Mathematical Monthly* (referred to as *AMM*) back to 1918. For the earlier *Gazettes* I used the index in Vol. 15 (1931), and for the earlier *Monthlies* (1918-50) I used Howard Eves and E. P. Starke, *The Otto Dunkel Memorial Problem Book.* These were godsends.

Everything in these works relating to my subject-matter and deemed worthy of inclusion has been included. I have also incorporated everything from my own prior publications, those containing the main results being found in *Australian Mathematics Teacher,* **7,** 7-10 (1951), **9,** 17-21 and 64 (1953), **16,** 64-5 (1960); *MG,* **45,** 94-7 (1961).

A name in parentheses is the originator's, if he is known for sure. If the references are to works only by Loyd or only by Dudeney, we can presume that the sole author referred to was the originator. But as Loyd and Dudeney used to exchange problems, references to works by both leave us uncertain.

Sometimes the dissection in the source is not quite the same as the one illustrated in this book. But it will be easily recognizable as a variant.

Ch. 1. The proof of plane dissectability is attributed to W. Bolyai and P. Gerwien in *Elemente der Mathematik,* **6,** 97 (1951), but no reference is given.

Fig. 1.6, *CP,* 84; 1.7, *PM,* 178; 1.9, *AMM,* **37**, 158-9 (1930), problems 3028 and 3048 (Harry C. Bradley); 1.16, *AM,* 154, *SL*2, 118; 1.17, *CP,* 77; 1.18, *PCP,* 195; 1.19, *AM,* 170, *SL*1, 79; 1.20, *MP,* 105.

Fig. 3.1, *AM,* 155 (similar); 3.2, *MP,* 108; 3.9, unpublished (similar, Bruce R. Gilson).

Fig. 4.8, *AMM,* **59**, 106-7 (1952), problem E972 (Michael Goldberg); 4.12, *SL*2, 80.

Fig. 5.2, *CP,* 26; 5.6, unpublished (Irving L. Freese).

Fig. 8.6, *MG,* **8**, 112 (1915), Fig. 26 (W. H. Macaulay).

Ch. 8. The conditions triangles must satisfy for four-piece, five-piece, and six-piece dissections were investigated by W. H. Macaulay, *MG,* **7**, 381-8 (1914), **8**, 72-6 and 109-15 (1915). They are rather intricate and, believing that this matter would interest very few readers, I have given no details.

Fig. 9.1, *AM,* Figs. 8 and 9 (originator not named); 9.2, *MP,* 103; 9.3, *AM,* Figs. 25 and 26, *CP,* 19, *SL*2, 128; 9.4, *PCP,* 182; 9.5, *AM,* 149; 9.15, *MP,* 112, *PM,* 180, *SL*2, 79 (similar); 9.17, W. W. Rouse Ball, *Mathematical Recreations and Essays,* 11th ed. p. 88 (Henry Perigal).

Figs. 10.1 and 10.3, *PCP,* 180.

Fig. 11.1, *AM,* 151 and 152, *SL*1, 100, *SL*2, 80 and 109; 11.5, *MP,* 111 (A. E. Hill); 11.7, *AM,* Figs. 37 and 38, *SL*2, 34; 11.8, *AM,* Figs. 31 and 32.

Fig. 12.1, Ball's *Recreations,* p. 92 (James Travers); 12.2, unpublished (Freese); 12.3, *AM,* Figs. 42 and 43; 12.4, *MP,* 110.

Fig. 14.6, *CP,* 8; 14.7, *SL*2, 115; 14.9, *SL*1, 11.

Figs. 16.10 to 16.13, unpublished (Freese).

Fig. 18.9, *PCP,* 184; 18.23, *Scientific American,* November 1961, pp. 168-9, December 1961, p. 158 (rectangle by Martin Gardner); 18.24, *Recreational Mathematics Magazine,* June 1961, p. 45 (Joseph S. Madachy).

Fig. 21.5, unpublished (C. Dudley Langford); 21.6, see text; 21.10, *AM,* 143, *SL*1, 20, *SL*2, 93; 21.14, H. Steinhaus, *Mathematical Snapshots,* 2nd ed., p. 8 (the reference on p. 256 to *AM,*

p. 27 is an error); 21.15, suggested by Joseph Rosenbaum's 13-piece dissection in *AMM,* **54,** 44 (1947), problem E721; 21.16, *MG,* **44,** 107 (1960) (Langford); 21.19, *MG,* **44,** 46 (1960) (Langford); 21.24, *MG,* **40,** 218 (1956) (Langford); 21.25, *MG,* **44,** 109 (1960) (Langford); 21.27 and 21.28, unpublished (Freese).

Fig. 22.9, *SL2,* 147; 22.10, *AM,* 156.

Fig. 23.1, suggested by CP, 37 and *SL2,* 18 (the six-piece dissection in *SL2,* 18 is only approximate); 23.3, *SL2,* 51; 23.9, *SL2,* 60; 23.10, *PCP,* 183 (Loyd); 23.11, *AM,* 160, *SL2,* 62; 23.12, *PCP,* 187, *SL1,* 59.

Figs. 24.1-24.3, *Eureka,* **14,** 23 (1951) (R. F. Wheeler); 24.4, *AMM,* **42,** 509 (1935), problem E4 (A. H. Wheeler); dissection of zonohedron, *AMM,* **56,** 694 (1949), problem E860 (Leo Moser); Figs. 24.11 and 24.12, *Comm. Math. Helv.,* **24,** 204-18 (1950) (J. P. Sydler).

Fig. D1, *AM,* 153; D2, *AM,* 150; D4, *AMM,* **28,** 186-7 (1921), problem 2799 (Harry C. Bradley, attributed in *MP,* 109 to E. B. Escott because his was the solution published); D51 and D52, unpublished (Freese); D55, *New Scientist* No. 224, 2 March 1961, p. 560 (Thomas H. O'Beirne), and for further interesting solid dissections see his *Puzzles and Paradoxes* (Oxford University Press).

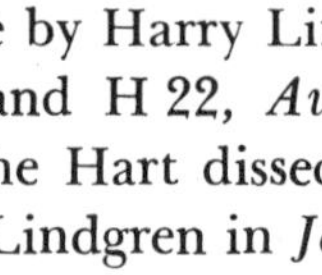

SL2 144

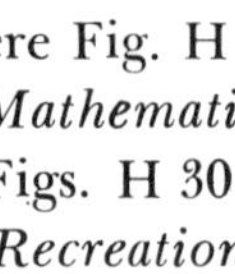

Sources for Appendix H

Dissections done by Harry Lindgren were Fig. H 6, unpublished, and Figs. H 1 and H 22, *Australian Mathematics Teacher,* **20,** 52-4 (1964). The Hart dissections in Figs. H 30 and H 31 were republished by Lindgren in *Journal of Recreational Mathematics,* **2,** 178-80 (1969). These were taken from *Matematica Dilettevole e Curiosa* by Italo Ghersi, reprinted by Hoepli of Milan in 1963. Fig. H 32 is to be published as a problem by James Schmerl in *Journal of Recreational Mathematics,* **5,** No. 4.

All other material is by Greg Frederickson. Some of this has been published in *Journal of Recreational Mathematics,* **5,** 22-26, 128-132 (January and April 1972). More is included in the July 1972 issue (as yet unpaged), and the rest will appear in later issues of *JRM.*

Appendix G

RECENT PROGRESS

Six of the dissections in Chapter 17 and two in Chapter 20 were improved on while the manuscript of this book was with the printers. For this reason, it was more convenient to include the improvements by adding a further appendix, rather than by altering the main text. The corresponding entries in Appendix E have been altered to suit.

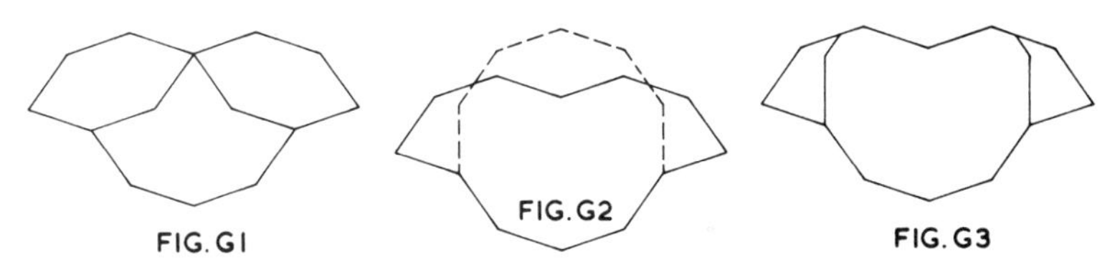

FIG. G1 FIG. G2 FIG. G3

The new decagon dissections are all based on a single tessellation, whose element has the same outline as that of Fig. B10. The element of the latter is repeated in Fig. G1 for comparison with the new one, which was found by superposing an undissected decagon on the outline of Fig. G1, as in Fig. G2. The new element, shown in Fig. G3, has a larger large piece than Fig. G1; indeed, it is clear from Fig. G2 that this element has the largest large piece possible. The strips derived from its tessellation will likewise have a larger large piece than those in Figs. 17.1-17.3. Thus, I have belatedly followed my own precept as laid down in Chapter 3 (Figs. 3.5 and 3.6).

The dissection into a pentagon in Fig. G4 is obtained by the method of Chapter 13, whereby we find the most favorable pair of vertical congruent lines on the tessellation. Dissections into triangle, golden rectangle, octagon, hexagram, and Greek cross are given in Figs. G5-G10. They all use the same decagon strip, ob-

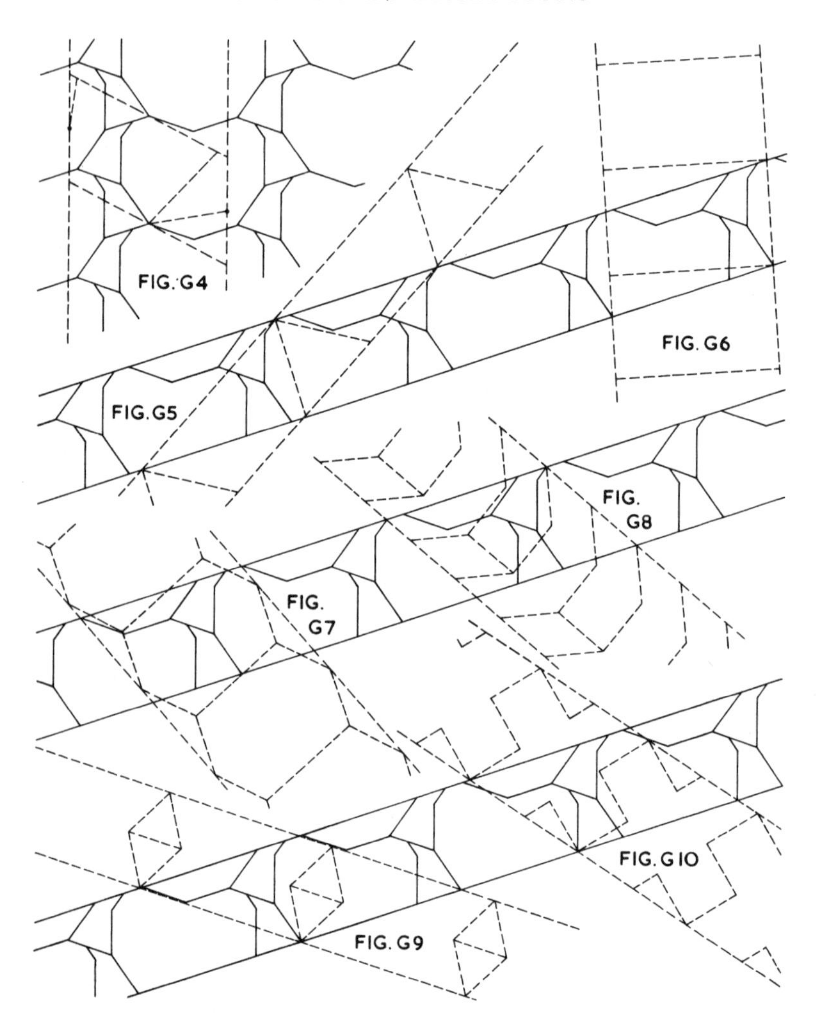

tained by drawing congruent lines on the tessellation that are inclined at 18° to the horizontal.

In Figs. G5, G7, G9, and G10, there are places where cuts almost cross. If they did, more pieces would be required. It has of course been verified by calculation that the critical places are as drawn.

Figures G11-G13 show three ways of dissecting {12/2} and {6/2} in nine pieces, with no pieces turned over as in Fig. 20.12. The first step for Fig. G11 is to rearrange the {12/2} as on the left. This rearrangement is the element of a tessellation with the same period

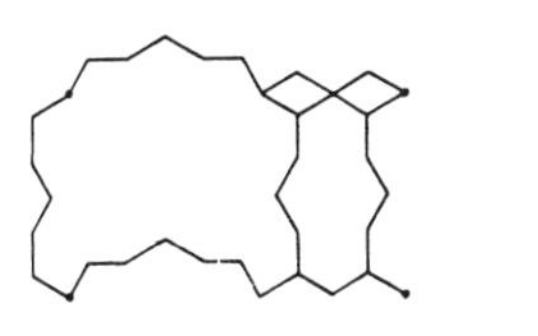
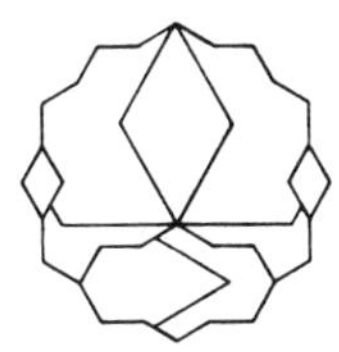

FIG. G11

parallelogram as Fig. B14, and the dissection is found by superposing the tessellations.

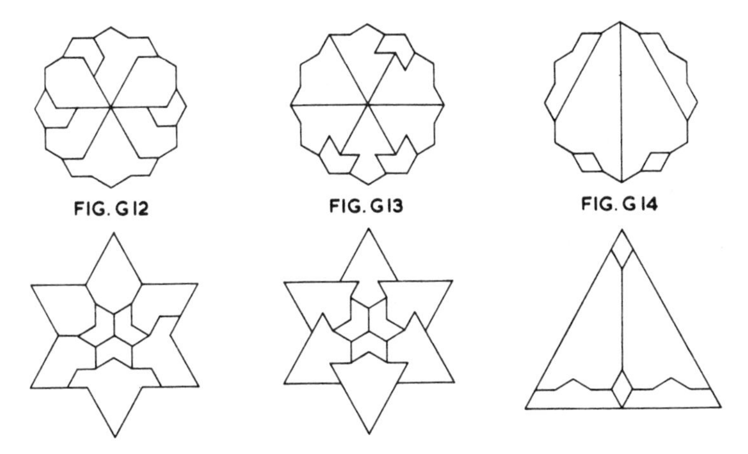

FIG. G12 FIG. G13 FIG. G14

The dissection of {12/2} and triangle in Fig. G14 is an improvement on Fig. 20.25, in that it has only six pieces instead of nine—a drastic reduction. How strikingly this bears out the statement in Chapter 1, that the subject is nowhere near exhaustion!

The decagon now seems to have been adequately dealt with, since further improvements will probably only be marginal. It is otherwise with the nine-gon. What we need for it is a rearrangement into a tessellation element as useful as Fig. G3, providing a breakthrough. Meanwhile the unsatisfactory state of nine-gon dissections gives no grounds for complacency.

Appendix H

EIGHT YEARS AFTER

As Harry Lindgren emphasized in Appendix G, the subject of geometric dissections is nowhere near exhaustion. Since he wrote that statement, thirty-one more dissections have been found, some of them new and some of them improvements on dissections already shown in this book. The purpose of this appendix is to present these dissections, which have been found in the eight years since the first publication of this work. Sources for all the dissections given herein are indicated on p. 162. Most of the dissections in the appendix are my work, and they reflect my interest in star dissections and all types of assemblies. If some readers feel that I have worked these areas to death, they can find solace in Figs. H 30, H 31, and H 32, which are by all accounts quite remarkable. Fig. H 32 illustrates the fact that there will always be a dissection worth discovering, if only someone ingenious enough will take the time to discover it. As for the organization of this appendix, the dissections which follow are arranged in the order in which they would have appeared in the text.

A Strip Dissection. The immediately preceding appendix, Appendix G, was written by Harry Lindgren in order to include improvements on some of his dissections after the main text had been sent to the printers. A number of improvements on decagon dissections just made it into the book under those circumstances. However, there is one dissection which just missed being included in Appendix G. It is a thirteen-piece dissection of a decagon and a heptagon, shown in Fig. H 1. Using the same decagon strip as in Figs. G 5-G 10, it betters the dissection in Fig. 17.5 by one piece.

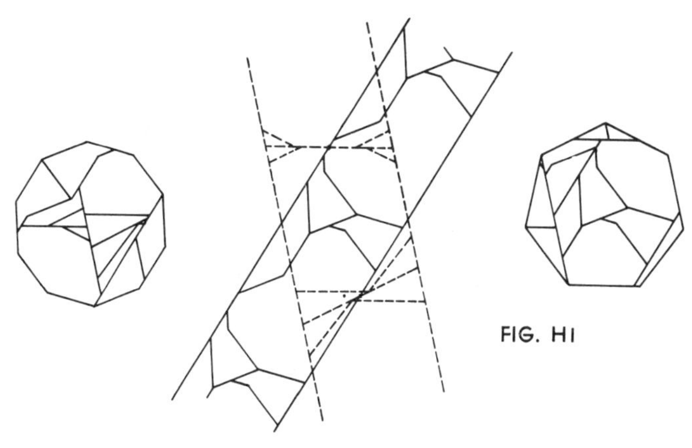

FIG. H1

Star Dissections. As always the prime concern in working dissections is finding a minimal solution. With this in mind, four dissections will be presented which are improvements on dissections found in Chapters 19 and 20. Three of these involve a saving of only one piece, so I cannot claim any great innovative ability. However, these improvements will confirm the adage that two heads are better than one. In addition to these improvements, there will be a number of new dissections and several generalizations accompanying them.

The first improvement comes on the eight-piece dissection of {8/2} into {4}, shown in Fig. 19.6. In this figure, four of the star's corners were cut off in the process of getting a tessellation. In the improvement on Fig. 19.6, the corners were left attached while the cuts signified by the dotted lines were made. The resulting pieces were then fitted into the shape of a square, and it was noted where there was overlap and holes. By trial and error it was determined how to cut out the overlap and fill in the holes. The key to the dissection is the jagged piece, and the seven-piece solution making use of this piece is shown in Fig. H 2.

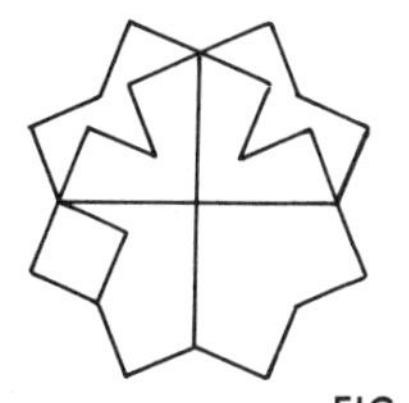

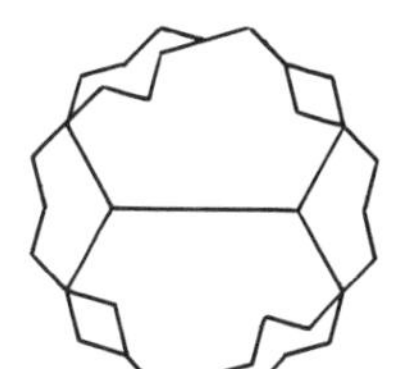

FIG. H2

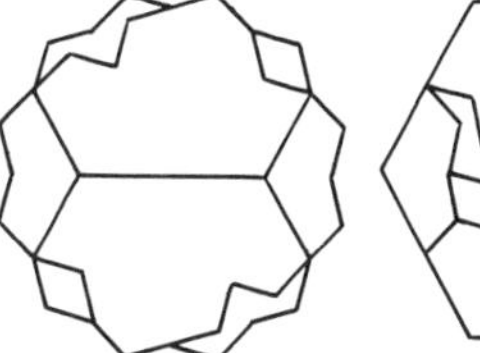

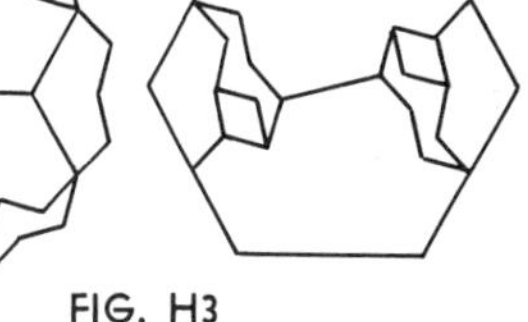

FIG. H3

The next improvement comes on the ten-piece dissection of {12/2} into {6}, shown in Fig. 20.4. The dissection in Fig. 20.4 is similar to the dissection in Fig. 19.6, in that some of the corners of the star were cut off in order to get a tessellation. But since Fig. 19.6 can be improved on, it would seem to follow that Fig. 20.4 could be improved on in much the same way. So it could be, and the resulting eight-piece dissection is shown in Fig. H 3. Note that there are again the jagged pieces, although they are slightly flattened in shape.

The next two improvements deal with {12/5} into {12}, shown in Fig. 20.6, and {12/5} into {4}, shown in Fig. 20.22. Both dissections are impressive as they stand; and yet they are not minimal. They can be improved by cutting sections off certain pieces and adding them onto other pieces: a three-way trade in the first dissection, and a four-way trade in the second. This "musical chairs" approach to manipulating the pieces resulted in a one-piece saving in each dissection. The two new dissections are shown in Figs. H 4 and H 5.

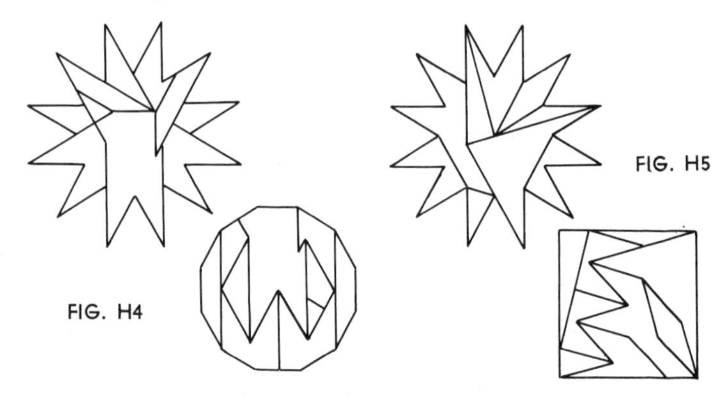
FIG. H4

FIG. H5

A fifth star dissection, Problem 45 in Appendix A, was also improved upon. This is an eight-piece dissection of {8/2} into a Greek cross, whose solution is shown in Fig. D 45. Since {8/2} is dissectable into a square via tessellations, it is also dissectable into a Greek cross in the same way. Harry Lindgren took the tessellation element gained from Fig. H 2, superimposed it over the tessellation of Greek crosses, and found the seven-piece dissection shown in Fig. H 6. Note that there are some very sharp points, the angles of which are just over 4°.

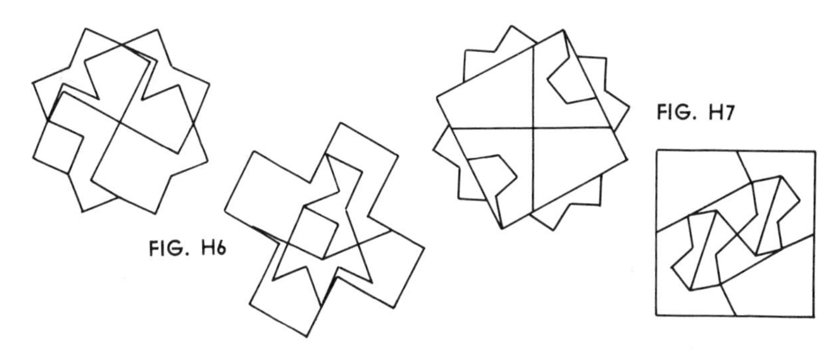

FIG. H6

FIG. H7

As yet no RTF dissections have been found for either $\{12/3\}$ or $\{12/4\}$, and it is quite likely that there are none with figures of less than twenty-four points. However, a dissection of $\{12/4\}$ into $\{4\}$, not based on an RTF, is possible and is shown in Fig. H 7. The dissection uses much the same tessellation procedure as that used in Fig. 19.6. The corners are cut off the star, leaving a square, and are arranged into a tessellation so that they could be cut into a smaller square. Cutting them into a square and combining it with the larger square would yield a twelve-piece solution. But the corners need not be cut at all if holes are cut for them to fit into. Doing this yields the ten-piece result shown in Fig. H 7.

In Chapter 20 Harry Lindgren lists the possible dissections among the seven-pointed and fourteen-pointed stars, but he does not perform any of the dissections. This may be because several of the dissections are very similar to those already done for the five-pointed and ten-pointed stars. For example, the dissection of $\{14/5\}$ into two $\{7/3\}$'s is very similar to that of $\{10/4\}$ into two $\{5/2\}$'s, and it thus follows the pattern of Fig. 20.20. In addition the three dissections relating $\{14/4\}$, $\{14/6\}$, and two $\{14\}$'s follow the pattern of Figs. 20.13 and 20.16, which relate $\{10/3\}$, $\{10/4\}$, and two $\{10\}$'s.

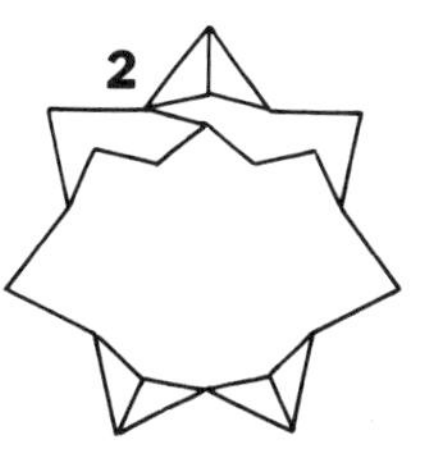

FIG. H8

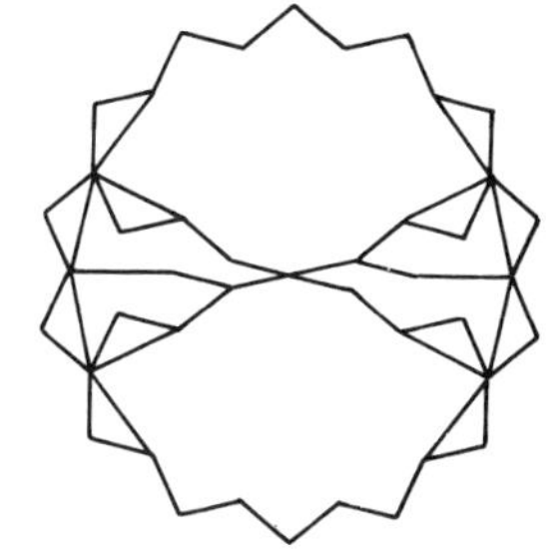

However, the remaining dissections in the group do not follow this type of pattern, and they call for original approaches. Fig. H 8 shows an eighteen-piece dissection of one $\{14/3\}$ into two $\{7/2\}$'s, and Fig. H 9 shows a fourteen-piece dissection of $\{14/2\}$ into $\{7\}$. While I am not at all convinced that this last dissection is minimal, efforts to produce a dissection along the lines of Fig. 20.3 have yielded at best a fifteen-piece solution. In any case Fig. H 9 does give an attractively symmetric dissection which can be generalized. For all dissections of $\{2n/2\}$ and $\{n\}$ where n is odd, a similar dissection is possible, and the number of pieces will be $2n$.

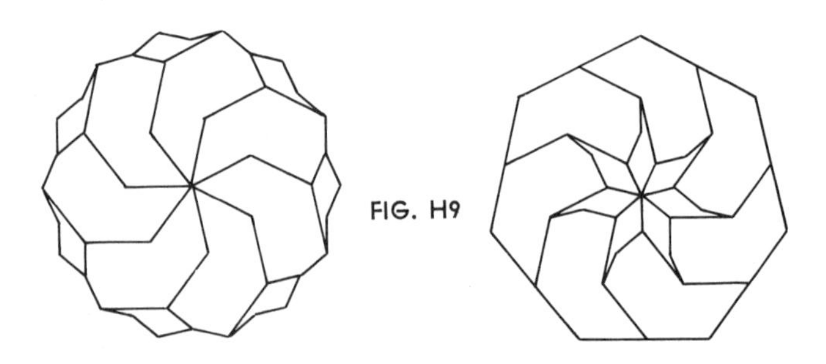
FIG. H9

At this point the reader may notice that all the seven-pointed and fourteen-pointed stars are accounted for, i.e., each has been included in an RTF dissection, just as Harry Lindgren had described in Chapter 20. But the surprising fact is that there are additional relations among these stars which Harry Lindgren did not discover. The key to these relations is

$$s\{14/5\}/s\{14/2\} = 2\cos(\pi/7)$$

This yields a fourteen-piece dissection as shown in Fig. H 10. The above line comes from the generalization

$$s\{2n/\tfrac{1}{3}(2n+1)\}/s\{2n/\tfrac{1}{3}(n-1)\} = 2\cos\{\pi(n-1)/6n\}$$

Substituting $n = 4$ will give again the dissection of $\{8/3\}$ into $\{8\}$, and substituting $n = 10$ will predict a dissection of $\{20/7\}$ into $\{20/3\}$.

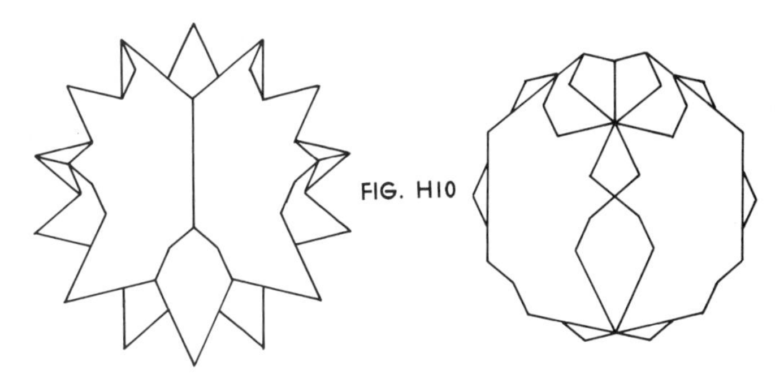

The relation between {14/5} and {14/2} is the link by which several other relations between seven-pointed and fourteen-pointed stars are made possible. These are

$$s\{7/3\}/s\{14/2\} = \sqrt{2} \cdot \cos(3\pi/14)/\sin(\pi/14),$$
$$s\{7/3\}/s\{7\} = \sqrt{2} \cdot 2\cos(\pi/7),$$
$$s\{7\}/s\{14/5\} = \cos(\pi/14)/\cos(\pi/7).$$

No generalization has been found for any of these special relations.

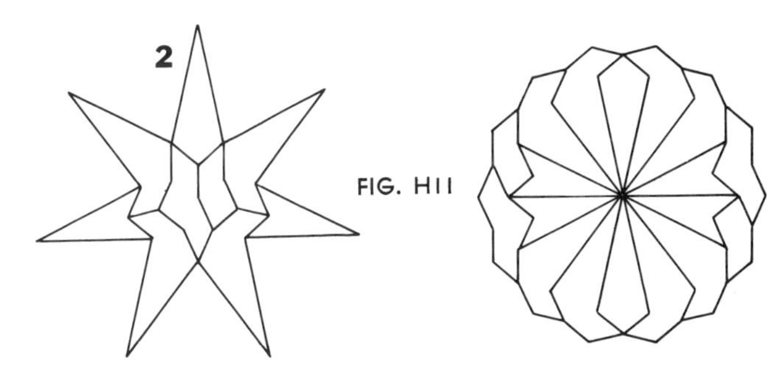

The first relation indicates a dissection of one {14/2} into two {7/3}'s, and the eighteen-piece dissection is shown in Fig. H 11. The second relation indicates that there should be dissections of one {7/3} into two {7}'s, and also of two into one. The first of these, one {7/3} into two {7}'s, is by far the most remarkable, requiring only nine pieces, as shown in Fig. H 12. After this dissection was found, a most amazing comparison between Figs. H 11 and H 12 was discovered. Each piece and each cut in one of the {7/3}'s has an analogous piece or cut in the other {7/3}. Of course there is also the dissection of two {7/3}'s into one

{7}. The dissection in Fig. H 13 requires fourteen pieces, and it is curious to note that one of the {7/3}'s has the same cuts as in Fig. H 12. Finally, the third relation predicts a dissection of {14/5} into {7}, but a simple one has not yet been found.

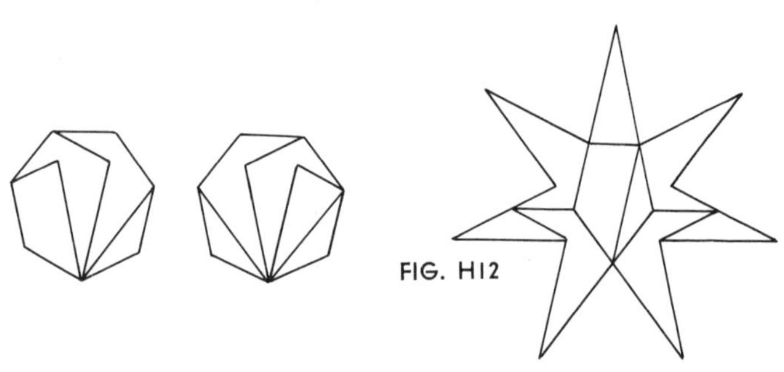

FIG. H12

The reader has probably noticed that there are in theory an infinite number of RTF dissections dealing with stars. However, as the dissection possibilities are exhausted for the simpler cases, the dissectionist must resort to stars with larger numbers of points and thus generally more pieces. Some of the dissections in the last group may already have seemed to require too many pieces for some readers' tastes. Yet one is led on in his search for more dissections by two goals. First is the desire to find (accidentally perhaps) dissections requiring unusually few pieces, such as in one {7/3} into two {7}'s. The second goal is to discover some overall general relation between these RTF dissections. While there are individual generalizations which lead in this direction, there is no unified pattern to these generalizations, so that a really adequate explanation is still lacking.

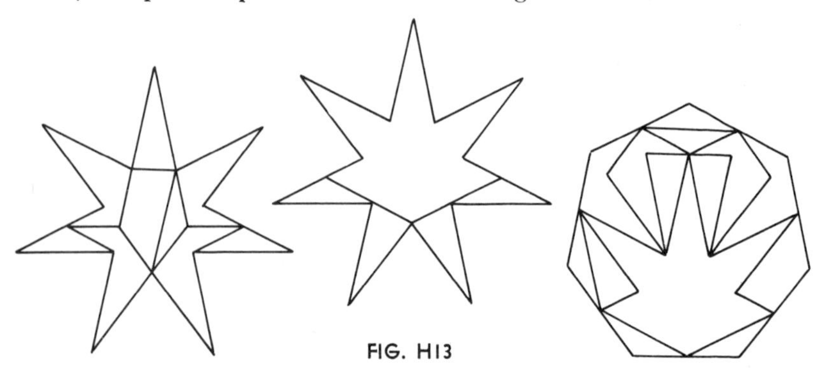

FIG. H13

The nine-pointed stars were investigated with respect to the first goal, and several interesting relations were discovered. First of all, {9/3} can be cut up into pieces which can be rearranged into a tessellation element. The tessellation repeats itself in a hexagonal pattern, with the element coming in groups of three. Fig. H 14 shows the star cut into the tessellation element, and beside it the nine-piece dissection of {9/3} into {6}. The trigonometric relation describing the dissection is

$$s\{6\}/s\{9/3\} = 2\cos(\pi/18)$$

As yet no generalization has been found for this relation.

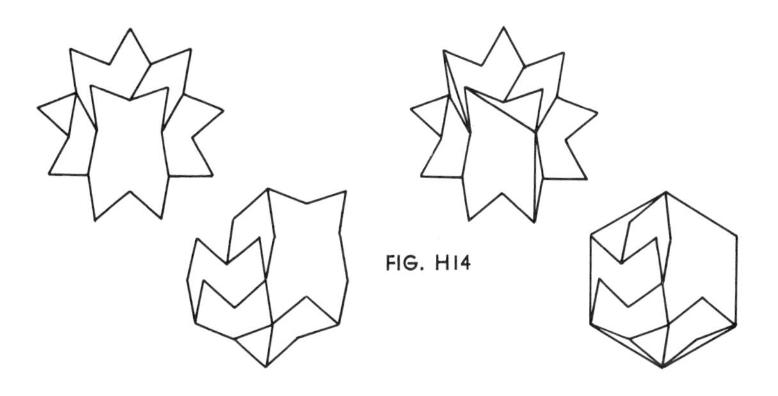

FIG. H14

Another unexpected relation was found between {9/2} and {9/4}. Both stars can be constructed by adding pieces onto a {9}. A symmetric nineteen-piece dissection is possible which looks something like the dissection between {10/3} and {10/4}, only more complicated. With a little work a fourteen-piece dissection such as Fig. H 15 can be discovered, but this may well not be minimal. The dissection is the special case of the general relation

$$s\{n/{}^1\!/_6\,(n+3)\}/s\{n/{}^1\!/_2\,(n-1)\} = 2\cos\,(\pi\,(n-1)\,/2n),$$

where n must be an odd multiple of three. If $n = 15$, then the relation predicts a dissection of {15/3} into {15/7}.

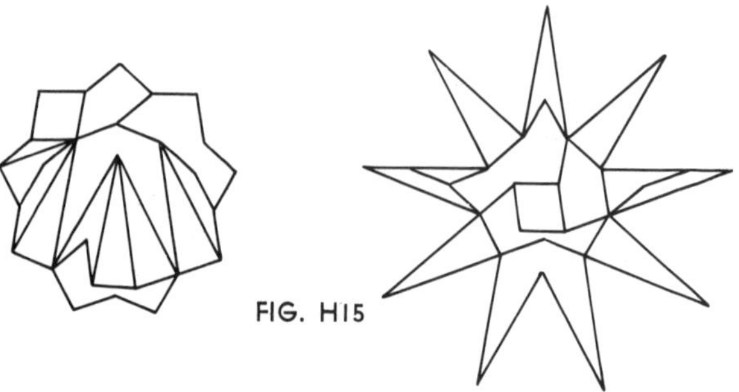
FIG. H15

Taking the above general relation and combining it with (12) from p. 95 yields the generalization

$$s\{n/\ {}^{1}\!/_{6}\ (n+3)\}/r\{2n/n-2\} = \sqrt{2}\cdot\cos(\pi(n-1)/2n),$$

where again n must be an odd multiple of three. With $n = 9$ the relation predicts an economical dissection of one $\{18/7\}$ into two $\{9/2\}$'s. The dissection is shown in Fig. H 16 and requires eighteen pieces. Note that each of the points of the $\{18/7\}$ is one ninth of a $\{9\}$, which is the reason why the dissection can be done so economically.

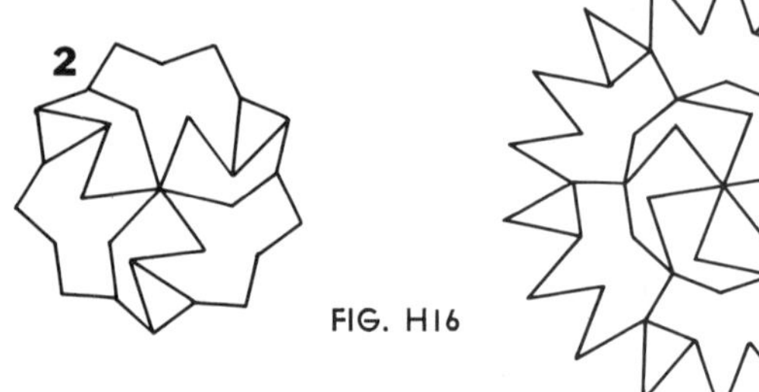

FIG. H16

Polygon Assemblies. Two of the polygon assemblies from Chapter 21 have been improved upon, and three new ones have been found. The first, shown in Fig. H 17, is a two into one assembly illustrating the case $n = 16$. It uses the same technique as the assembly for $n = 8$ in Fig. 21.17 and the assembly for $n = 4$ shown in Fig. 21.23. The polygons must be imagined to be divided up into rhombs, some of which will be squares. Each of the squares is cut in half along one of its diagonals, and then additional cuts are needed to rearrange some of the rhombs. This technique will give a two into one assembly in n pieces, which seems to be minimal in all cases except for $n = 12$.

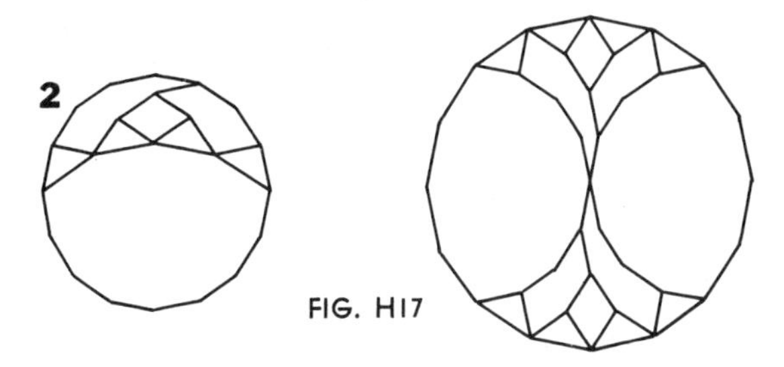

FIG. H17

Several interesting discoveries have been made about three into one assemblies. Fig. 21.19 depicts a twenty-one piece assembly for $n = 9$, which has been greatly improved upon. With a relatively minor rearrangement of cuts, an eighteen-piece assembly is possible. But the real surprise is that the number of pieces can be further reduced to fifteen, as shown in Fig. H 18. The trick is to imagine the small nine-gons cut up into rhombs and equilateral triangles. Three of the equilateral triangles do not have to be cut up or detached at all. Instead, holes were cut for these triangles to fit into. Doing this yielded an assembly that was not strictly symmetric, but there is a curious three-way balance in the way the small nine-gons are cut up.

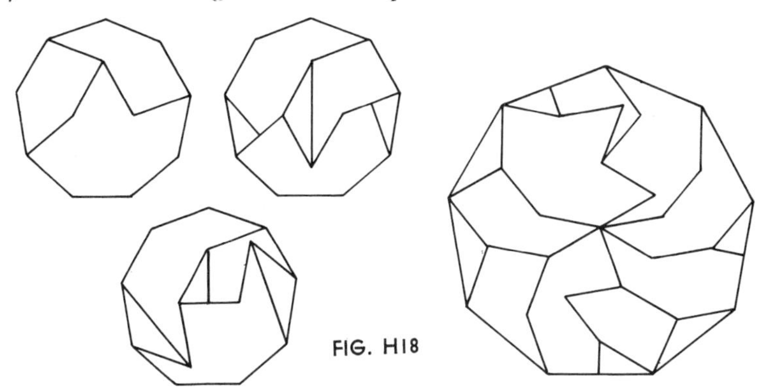
FIG. H18

Can the same trick used in the case of $n = 9$ be used for the case $n = 15$? The answer is yes. An obvious thirty-piece solution can be reduced to a twenty-four piece assembly, as shown in Fig. H 19. Once this assembly was achieved, a pattern for three into one assemblies became clear. The polygons which qualify for three into one assemblies can be divided into two groups. The first is composed of odd multiples of 3, such as $\{3\}$, $\{9\}$, $\{15\}$, and so on. These are represented by $n = 6p - 3$, where p is an integer. The second group is composed of even multiples of 3, such as {6}, {12}, {18}, and so on, and these are represented by $n = 6p$. In both cases the minimum number of pieces needed for the assembly seems to be $9p - 3$, although it has not yet been proven why this is the case.

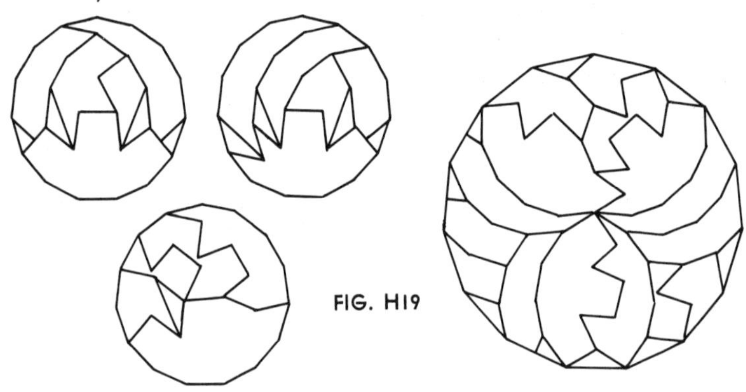

FIG. H19

In the category of five into one assemblies, the assembly of five decagons into one, shown in Fig. 21.25, has been improved on. While it is attractive to show the assembly in terms of five pie-shaped parts, in this case it hampers the finding of a minimal assembly. The trick used is to find a single cut in the small decagons which will produce the correct side-length for the large decagon. Each of the small decagons in Fig. H 20 has this cut down the center, and an eighteen-piece assembly results. Although it was not discovered until after the fact, note the similarity of the pieces in the assembly with those in the decagon strip shown in Fig. 17.1.

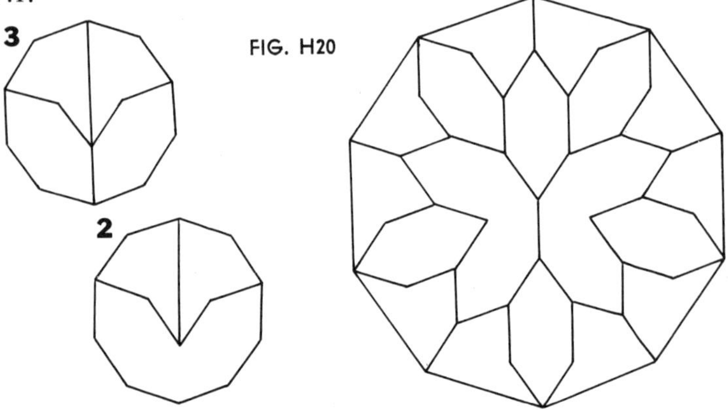

FIG. H20

The last polygon assembly in this section is that of six dodecagons into one. While Harry Lindgren mentioned that this assembly is possible, he did not attempt to find it. He should not have passed over it, for it turned out to be a very stimulating challenge. With a lot of effort the number of pieces can be whittled down to twenty-four, as shown in Fig. H 21. The important trick is to find a single cut which will produce the side-length of the large dodecagon. However, to get the assembly down to twenty-four pieces, a number of small tricks must be orchestrated together. Each one adds a little more asymmetry, and the assembly turns out to be one of the most asymmetrical of the dissections depending on RTF's.

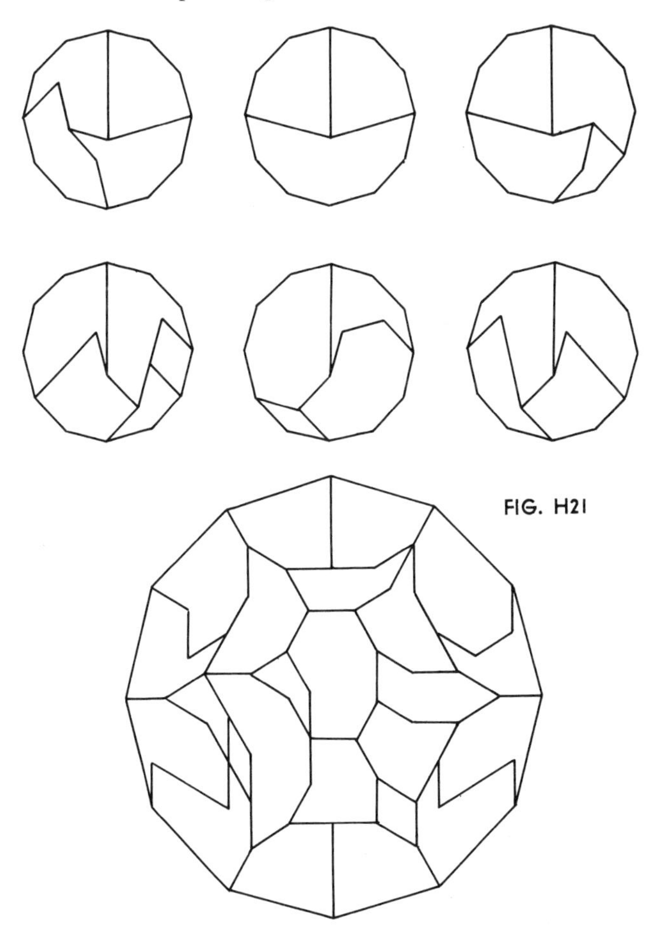

FIG. H21

Star Assemblies. Both of the star assemblies shown in Chapter 21 have been improved upon. The first, three {6/2}'s into one {6/2} shown in Fig. 21.22, was improved by Harry Lindgren shortly after the book was published. The new assembly is shown in Fig. H 22 and requires only twelve pieces. The trick needed to reduce the number of pieces from thirteen to twelve is that each of the three small stars must be cut up.

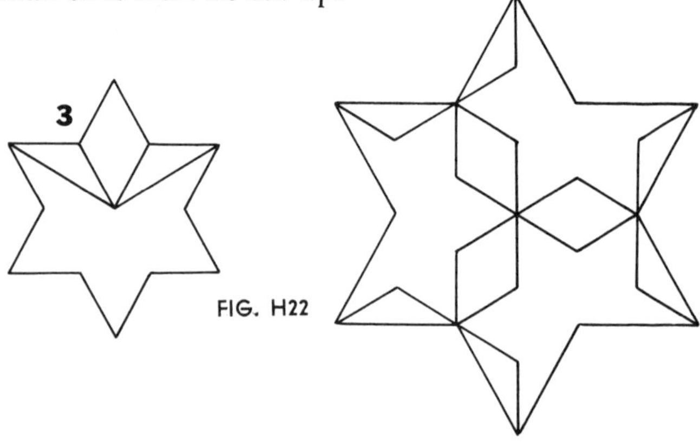

FIG. H22

There are actually any number of different variations of this assembly, each requiring only twelve pieces. My own favorite is an original assembly shown in Fig. H 23. Fig. H 23 has been included because there are only two different shaped pieces in it and they are close to being the same size. In addition, the center of the large star is a hexagon and is derived from the assembly of two {6/2}'s into one {6}, shown in Fig. 20.8.

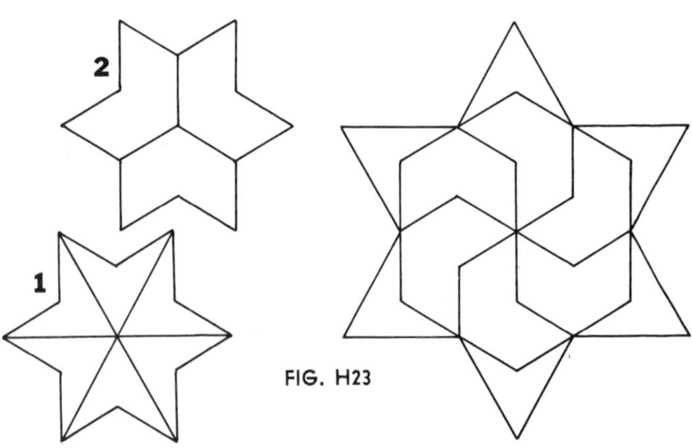

FIG. H23

The other star assembly presented in Chapter 21 is the sixteen-piece assembly of two {12/3}'s into one {12/3}, shown in Fig. 21.23. That assembly did not appear minimal because of four thin slivers, which seemed a blemish on the assembly. Thus the assembly was modified to avoid these, and the twelve-piece assembly of Fig. H 24 was the result. This assembly, like the one in Fig. 21.23, is based on the assembly of two squares, but the pieces which are analogous to the isosceles right triangles are more complicated. These winding pieces are like geometrical gerrymanders. They are extended to form as much of the outline of the large {12/3} as possible, and at the same time they interlock in the small {12/3}'s. More than likely, this assembly is minimal.

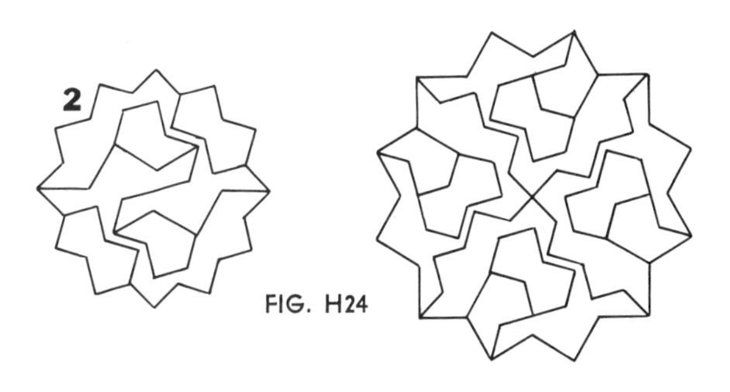

FIG. H24

The assembly in Fig. 21.23 was the only assembly that Harry Lindgren did with twelve-pointed stars. However, there are quite a few other assemblies which are possible. Besides {12/3}, the other twelve-pointed stars are {12/2}, {12/4}, and {12/5}. Not only do all of these qualify for two into one assemblies, but they also qualify for three into one assemblies. Thus there are eight possible assemblies, of which six appear in this section. The {12/2} is an easy prey for a two into one assembly. The star can be broken up into rhombs, as was done in the polygon assemblies. Some of these rhombs are squares, and of course the diagonal of the square will produce the correct side-length in a two into one assembly. The thirteen-piece assembly in Fig. H 25 can be arrived at without too much trouble.

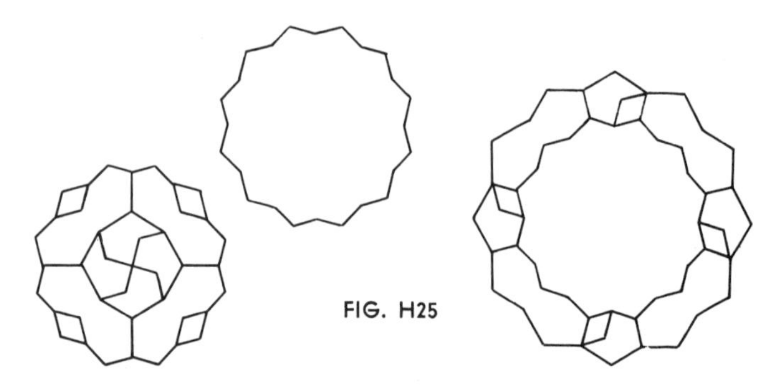

FIG. H25

The assembly of two {12/4}'s into one {12/4} is based on breaking the star up into rhombs and triangles. The {12/4} can be viewed as being made up of twenty-four equilateral triangles and twelve rhombs with 30° and 150° angles. While there are no squares in this structure, the equivalent of a square's diagonal is found in the line from the wide vertex of the rhombus to the far vertex of an adjacent triangle. The best solution found so far is the eighteen-piece assembly shown in Fig. H 26.

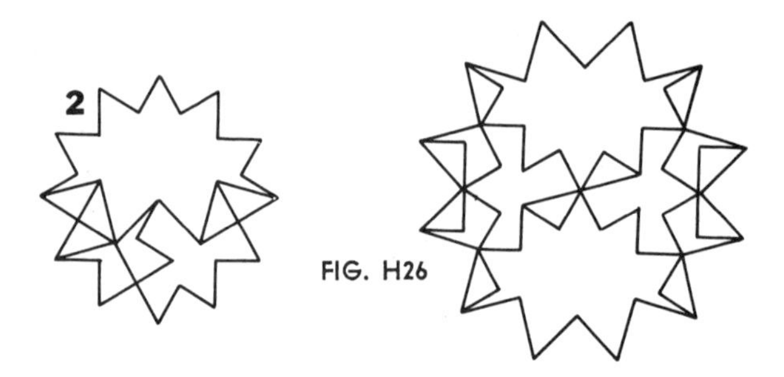

FIG. H26

The other assemblies presented in this section are three into one assemblies of {12/2}, {12/3}, and {12/4}. The assembly of three {12/2}'s was the easiest to discover, because it is based on a simple seven-piece assembly of three hexagons into one. In the seven-piece assembly two of the hexagons are cut into thirds and arranged around the remaining hexagon, which stays intact. In the {12/2} assembly, two of the small {12/2}'s are also cut into thirds, and are similarly arranged. However, the remaining {12/2} must be cut into twelve pieces to fill in the holes in the larger figure. The result is the eighteen-piece assembly shown in Fig. H 27.

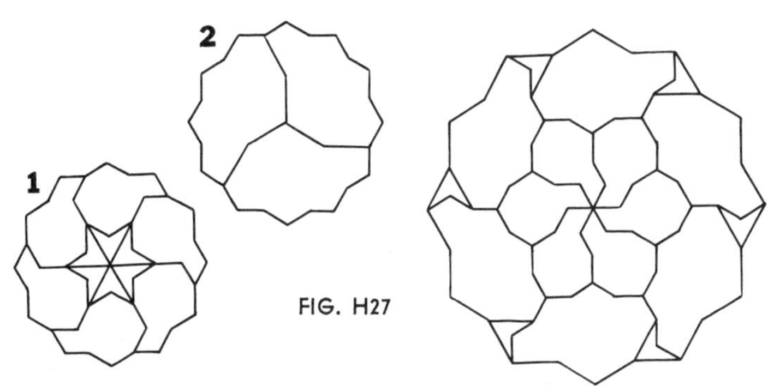

FIG. H27

The assembly of three {12/3}'s into one was much more difficult to find. It is partially based on breaking the figure up into rhombs, but another element must be taken into account. This is the jagged piece, taken from Fig. H 3. While not strictly reducible into rhombs, it is equal in area to a square plus a rhombus with 30° and 150° angles. The twenty-one piece solution shown in Fig. H 28 was even more difficult to discover because it does not have 60° rotational symmetry like Figs. H 27 and H 29, but rather 90° rotational symmetry.

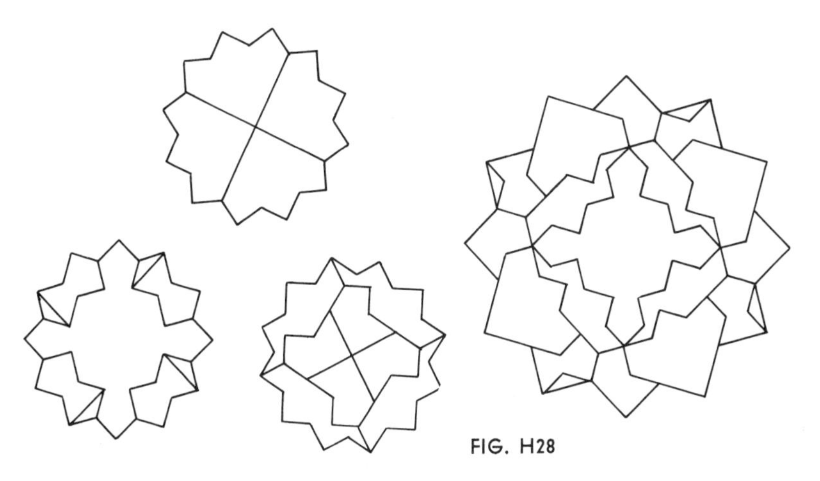
FIG. H28

The final star assembly is the twenty-four piece solution of three {12/4}'s into one shown in Fig. H 29. It can be found again by breaking the figure up into squares and equilateral triangles, with the modification that some of the squares will have equilateral triangles cut out of one of their sides. This assembly came about on an afternoon when I had planned to go to a

Baltimore Orioles–Oakland A's play-off game, but the game was rained out. Thus discovering the assembly was a form of consolation, and a very good form indeed!

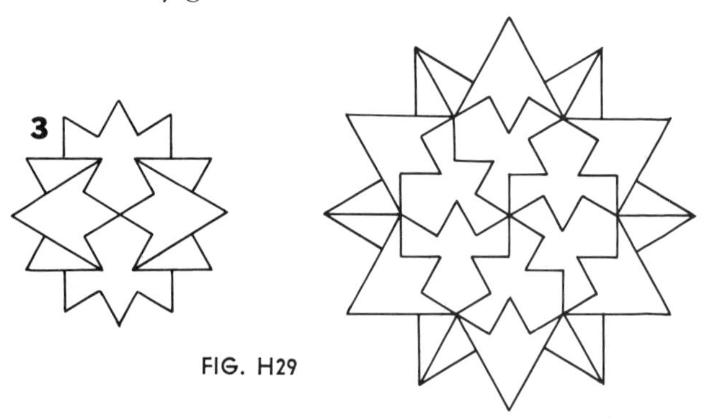

FIG. H29

If the reader looks carefully at the three into one assemblies of the twelve-pointed figures shown in Figs. 21.21, H 27, H 28, and H 29, he may be able to discern a pattern. Starting with the assembly of three $\{12\}$'s, each assembly requires just three more pieces than the preceding one. Why this is so has not yet been determined, but the hypothesis was quite helpful in discovering the twenty-one piece solution of three $\{12/3\}$'s after I had almost stopped with a twenty-four piece assembly. This little bit of hypothesizing also fits in nicely with what was done for three into one polygon assemblies, so that perhaps a proof of minimal numbers of pieces can follow from this.

Assemblies of Similar but Unequal Polygons. One of the numerous ingenious dissections due to Irving Freese is that of assembling two regular n-gons, whose relative size is immaterial, into one regular n-gon. The assembly requires $2n + 1$ pieces and is shown for a pentagon in Fig. 21.28. While this dissection is quite amazing in itself, it turns out to be a special case of two different generalizations. It is possible to do the same assembly on *any* polygon (regular or not) that has either an inscribed circle or a circumscribed circle. The assembly will again require just $2n + 1$ pieces. This more generalized problem was proposed and solved by Hart in 1877. It came to Harry Lindgren's attention less than a year after the publication of his book. Much of the description printed here is taken from the *JRM* article written by Lindgren.

Fig. H 30 shows the assembly of two pentagons which have an inscribed circle. In the larger of the two similar pentagons, perpendiculars are drawn from the center of the incircle to each of the sides. Each cut *BD* is made so that the tangent of the angle *DBC* is the ratio of any linear dimension in the smaller *n*-gon to the corresponding one in the larger. (This is precisely what was done in Fig. 21.28.) Since two tangents to a circle are equal, the five pieces corresponding to triangle *BCD* can be rotated around to fit onto side *BE*, as in the intermediate step. The pieces are then shifted outward in groups of two, so that the small pentagon will just fill in the hole in the center.

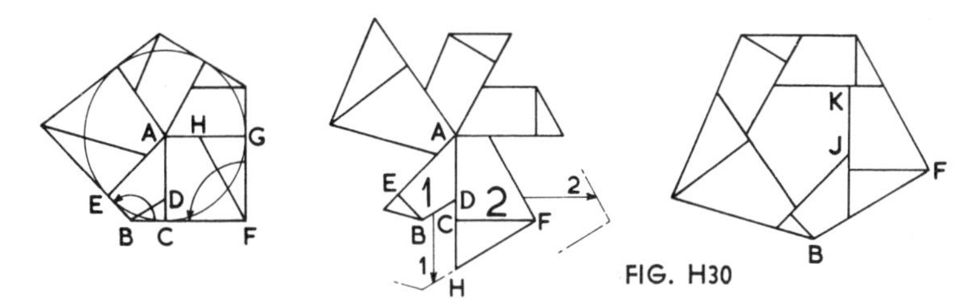

FIG. H30

The case of a polygon with a circumscribed circle is shown in Fig. H 31. In the larger pentagon, the lines drawn from the center of the circle must be the perpendicular bisectors of the sides. The cuts *BD* are located as they were in Fig. H 30. However, this time the five pieces corresponding to triangle *BCD* are rotated in the other direction. Thus *BC* fits onto side *CF* and the intermediate step is obtained. Finally, the pieces are again shifted outward, so that the smaller pentagon can be placed in the center.

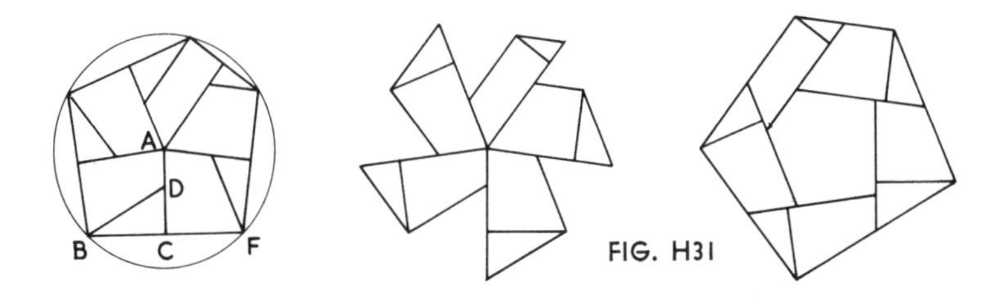

FIG. H31

The last assembly in this section is a special case of two hexagons into one, and it is remarkable in that it requires only five pieces. The sides of the hexagons are in a ratio of 3:4:5 and demonstrate the Pythagorean identity $3^2 + 4^2 = 5^2$. The dissection is a rational one, and is first cousin to Figs. 1.19 and 1.20, which deal with squares. In those figures each of the squares can be broken down into small squares with sides of unit length. In this dissection each of the hexagons can be broken down into equilateral triangles with sides of unit length. All cuts are made along the boundaries of these triangles, and the assembly in Fig. H 32 is the result.

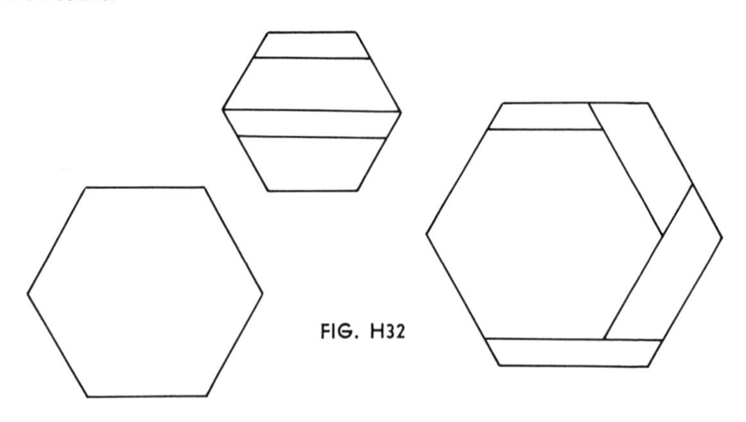

FIG. H32

This last assembly, originated by James Schmerl, was sent to me shortly before the manuscript was to be forwarded to the printer. Deceptively simple, it is in several ways one of the best dissections in Appendix H. First, it has the fewest number of pieces and thus offers a welcome relief from some of the more complicated dissections. Second, it extends dissection technique into a relatively undeveloped area. But finally, and most importantly, it demonstrates that no one has completely monopolized the field of geometric dissections, and that there is room for anyone who is willing to accept the challenge.

A CATALOGUE OF SELECTED DOVER BOOKS IN ALL FIELDS OF INTEREST

A CATALOGUE OF SELECTED DOVER BOOKS IN ALL FIELDS OF INTEREST

AMERICA'S OLD MASTERS, James T. Flexner. Four men emerged unexpectedly from provincial 18th century America to leadership in European art: Benjamin West, J. S. Copley, C. R. Peale, Gilbert Stuart. Brilliant coverage of lives and contributions. Revised, 1967 edition. 69 plates. 365pp. of text.
21806-6 Paperbound $3.00

FIRST FLOWERS OF OUR WILDERNESS: AMERICAN PAINTING, THE COLONIAL PERIOD, James T. Flexner. Painters, and regional painting traditions from earliest Colonial times up to the emergence of Copley, West and Peale Sr., Foster, Gustavus Hesselius, Feke, John Smibert and many anonymous painters in the primitive manner. Engaging presentation, with 162 illustrations. xxii + 368pp.
22180-6 Paperbound $3.50

THE LIGHT OF DISTANT SKIES: AMERICAN PAINTING, 1760-1835, James T. Flexner. The great generation of early American painters goes to Europe to learn and to teach: West, Copley, Gilbert Stuart and others. Allston, Trumbull, Morse; also contemporary American painters—primitives, derivatives, academics—who remained in America. 102 illustrations. xiii + 306pp. 22179-2 Paperbound $3.00

A HISTORY OF THE RISE AND PROGRESS OF THE ARTS OF DESIGN IN THE UNITED STATES, William Dunlap. Much the richest mine of information on early American painters, sculptors, architects, engravers, miniaturists, etc. The only source of information for scores of artists, the major primary source for many others. Unabridged reprint of rare original 1834 edition, with new introduction by James T. Flexner, and 394 new illustrations. Edited by Rita Weiss. 6⅝ x 9⅝.
21695-0, 21696-9, 21697-7 Three volumes, Paperbound $13.50

EPOCHS OF CHINESE AND JAPANESE ART, Ernest F. Fenollosa. From primitive Chinese art to the 20th century, thorough history, explanation of every important art period and form, including Japanese woodcuts; main stress on China and Japan, but Tibet, Korea also included. Still unexcelled for its detailed, rich coverage of cultural background, aesthetic elements, diffusion studies, particularly of the historical period. 2nd, 1913 edition. 242 illustrations. lii + 439pp. of text.
20364-6, 20365-4 Two volumes, Paperbound $6.00

THE GENTLE ART OF MAKING ENEMIES, James A. M. Whistler. Greatest wit of his day deflates Oscar Wilde, Ruskin, Swinburne; strikes back at inane critics, exhibitions, art journalism; aesthetics of impressionist revolution in most striking form. Highly readable classic by great painter. Reproduction of edition designed by Whistler. Introduction by Alfred Werner. xxxvi + 334pp.
21875-9 Paperbound $2.50

VISUAL ILLUSIONS: THEIR CAUSES, CHARACTERISTICS, AND APPLICATIONS, Matthew Luckiesh. Thorough description and discussion of optical illusion, geometric and perspective, particularly; size and shape distortions, illusions of color, of motion; natural illusions; use of illusion in art and magic, industry, etc. Most useful today with op art, also for classical art. Scores of effects illustrated. Introduction by William H. Ittleson. 100 illustrations. xxi + 252pp.
21530-X Paperbound $2.00

A HANDBOOK OF ANATOMY FOR ART STUDENTS, Arthur Thomson. Thorough, virtually exhaustive coverage of skeletal structure, musculature, etc. Full text, supplemented by anatomical diagrams and drawings and by photographs of undraped figures. Unique in its comparison of male and female forms, pointing out differences of contour, texture, form. 211 figures, 40 drawings, 86 photographs. xx + 459pp. 5⅜ x 8⅜. 21163-0 Paperbound $3.50

150 MASTERPIECES OF DRAWING, Selected by Anthony Toney. Full page reproductions of drawings from the early 16th to the end of the 18th century, all beautifully reproduced: Rembrandt, Michelangelo, Dürer, Fragonard, Urs, Graf, Wouwerman, many others. First-rate browsing book, model book for artists. xviii + 150pp. 8⅜ x 11¼. 21032-4 Paperbound $2.50

THE LATER WORK OF AUBREY BEARDSLEY, Aubrey Beardsley. Exotic, erotic, ironic masterpieces in full maturity: Comedy Ballet, Venus and Tannhauser, Pierrot, Lysistrata, Rape of the Lock, Savoy material, Ali Baba, Volpone, etc. This material revolutionized the art world, and is still powerful, fresh, brilliant. With *The Early Work,* all Beardsley's finest work. 174 plates, 2 in color. xiv + 176pp. 8⅛ x 11.
21817-1 Paperbound $3.00

DRAWINGS OF REMBRANDT, Rembrandt van Rijn. Complete reproduction of fabulously rare edition by Lippmann and Hofstede de Groot, completely reedited, updated, improved by Prof. Seymour Slive, Fogg Museum. Portraits, Biblical sketches, landscapes, Oriental types, nudes, episodes from classical mythology—All Rembrandt's fertile genius. Also selection of drawings by his pupils and followers. "Stunning volumes," *Saturday Review.* 550 illustrations. lxxviii + 552pp. 9⅛ x 12¼. 21485-0, 21486-9 Two volumes, Paperbound $10.00

THE DISASTERS OF WAR, Francisco Goya. One of the masterpieces of Western civilization—83 etchings that record Goya's shattering, bitter reaction to the Napoleonic war that swept through Spain after the insurrection of 1808 and to war in general. Reprint of the first edition, with three additional plates from Boston's Museum of Fine Arts. All plates facsimile size. Introduction by Philip Hofer, Fogg Museum. v + 97pp. 9⅜ x 8¼. 21872-4 Paperbound $2.00

GRAPHIC WORKS OF ODILON REDON. Largest collection of Redon's graphic works ever assembled: 172 lithographs, 28 etchings and engravings, 9 drawings. These include some of his most famous works. All the plates from *Odilon Redon: oeuvre graphique complet,* plus additional plates. New introduction and caption translations by Alfred Werner. 209 illustrations. xxvii + 209pp. 9⅛ x 12¼.
21966-8 Paperbound $4.00

DESIGN BY ACCIDENT; A BOOK OF "ACCIDENTAL EFFECTS" FOR ARTISTS AND DESIGNERS, James F. O'Brien. Create your own unique, striking, imaginative effects by "controlled accident" interaction of materials: paints and lacquers, oil and water based paints, splatter, crackling materials, shatter, similar items. Everything you do will be different; first book on this limitless art, so useful to both fine artist and commercial artist. Full instructions. 192 plates showing "accidents," 8 in color. viii + 215pp. 8⅜ x 11¼. 21942-9 Paperbound $3.50

THE BOOK OF SIGNS, Rudolf Koch. Famed German type designer draws 493 beautiful symbols: religious, mystical, alchemical, imperial, property marks, runes, etc. Remarkable fusion of traditional and modern. Good for suggestions of timelessness, smartness, modernity. Text. vi + 104pp. 6⅛ x 9¼.
20162-7 Paperbound $1.25

HISTORY OF INDIAN AND INDONESIAN ART, Ananda K. Coomaraswamy. An unabridged republication of one of the finest books by a great scholar in Eastern art. Rich in descriptive material, history, social backgrounds; Sunga reliefs, Rajput paintings, Gupta temples, Burmese frescoes, textiles, jewelry, sculpture, etc. 400 photos. viii + 423pp. 6⅜ x 9¾. 21436-2 Paperbound $4.00

PRIMITIVE ART, Franz Boas. America's foremost anthropologist surveys textiles, ceramics, woodcarving, basketry, metalwork, etc.; patterns, technology, creation of symbols, style origins. All areas of world, but very full on Northwest Coast Indians. More than 350 illustrations of baskets, boxes, totem poles, weapons, etc. 378 pp.
20025-6 Paperbound $3.00

THE GENTLEMAN AND CABINET MAKER'S DIRECTOR, Thomas Chippendale. Full reprint (third edition, 1762) of most influential furniture book of all time, by master cabinetmaker. 200 plates, illustrating chairs, sofas, mirrors, tables, cabinets, plus 24 photographs of surviving pieces. Biographical introduction by N. Bienenstock. vi + 249pp. 9⅞ x 12¾. 21601-2 Paperbound $4.00

AMERICAN ANTIQUE FURNITURE, Edgar G. Miller, Jr. The basic coverage of all American furniture before 1840. Individual chapters cover type of furniture—clocks, tables, sideboards, etc.—chronologically, with inexhaustible wealth of data. More than 2100 photographs, all identified, commented on. Essential to all early American collectors. Introduction by H. E. Keyes. vi + 1106pp. 7⅞ x 10¾.
21599-7, 21600-4 Two volumes, Paperbound $11.00

PENNSYLVANIA DUTCH AMERICAN FOLK ART, Henry J. Kauffman. 279 photos, 28 drawings of tulipware, Fraktur script, painted tinware, toys, flowered furniture, quilts, samplers, hex signs, house interiors, etc. Full descriptive text. Excellent for tourist, rewarding for designer, collector. Map. 146pp. 7⅞ x 10¾.
21205-X Paperbound $2.50

EARLY NEW ENGLAND GRAVESTONE RUBBINGS, Edmund V. Gillon, Jr. 43 photographs, 226 carefully reproduced rubbings show heavily symbolic, sometimes macabre early gravestones, up to early 19th century. Remarkable early American primitive art, occasionally strikingly beautiful; always powerful. Text. xxvi + 207pp. 8⅜ x 11¼. 21380-3 Paperbound $3.50

ALPHABETS AND ORNAMENTS, Ernst Lehner. Well-known pictorial source for decorative alphabets, script examples, cartouches, frames, decorative title pages, calligraphic initials, borders, similar material. 14th to 19th century, mostly European. Useful in almost any graphic arts designing, varied styles. 750 illustrations. 256pp. 7 x 10. 21905-4 Paperbound $4.00

PAINTING: A CREATIVE APPROACH, Norman Colquhoun. For the beginner simple guide provides an instructive approach to painting: major stumbling blocks for beginner; overcoming them, technical points; paints and pigments; oil painting; watercolor and other media and color. New section on "plastic" paints. Glossary. Formerly *Paint Your Own Pictures*. 221pp. 22000-1 Paperbound $1.75

THE ENJOYMENT AND USE OF COLOR, Walter Sargent. Explanation of the relations between colors themselves and between colors in nature and art, including hundreds of little-known facts about color values, intensities, effects of high and low illumination, complementary colors. Many practical hints for painters, references to great masters. 7 color plates, 29 illustrations. x + 274pp.
20944-X Paperbound $2.75

THE NOTEBOOKS OF LEONARDO DA VINCI, compiled and edited by Jean Paul Richter. 1566 extracts from original manuscripts reveal the full range of Leonardo's versatile genius: all his writings on painting, sculpture, architecture, anatomy, astronomy, geography, topography, physiology, mining, music, etc., in both Italian and English, with 186 plates of manuscript pages and more than 500 additional drawings. Includes studies for the Last Supper, the lost Sforza monument, and other works. Total of xlvii + 866pp. 7⅞ x 10¾.
22572-0, 22573-9 Two volumes, Paperbound $10.00

MONTGOMERY WARD CATALOGUE OF 1895. Tea gowns, yards of flannel and pillow-case lace, stereoscopes, books of gospel hymns, the New Improved Singer Sewing Machine, side saddles, milk skimmers, straight-edged razors, high-button shoes, spittoons, and on and on . . . listing some 25,000 items, practically all illustrated. Essential to the shoppers of the 1890's, it is our truest record of the spirit of the period. Unaltered reprint of Issue No. 57, Spring and Summer 1895. Introduction by Boris Emmet. Innumerable illustrations. xiii + 624pp. 8½ x 11⅝.
22377-9 Paperbound $6.95

THE CRYSTAL PALACE EXHIBITION ILLUSTRATED CATALOGUE (LONDON, 1851). One of the wonders of the modern world—the Crystal Palace Exhibition in which all the nations of the civilized world exhibited their achievements in the arts and sciences—presented in an equally important illustrated catalogue. More than 1700 items pictured with accompanying text—ceramics, textiles, cast-iron work, carpets, pianos, sleds, razors, wall-papers, billiard tables, beehives, silverware and hundreds of other artifacts—represent the focal point of Victorian culture in the Western World. Probably the largest collection of Victorian decorative art ever assembled—indispensable for antiquarians and designers. Unabridged republication of the Art-Journal Catalogue of the Great Exhibition of 1851, with all terminal essays. New introduction by John Gloag, F.S.A. xxxiv + 426pp. 9 x 12.
22503-8 Paperbound $4.50

A HISTORY OF COSTUME, Carl Köhler. Definitive history, based on surviving pieces of clothing primarily, and paintings, statues, etc. secondarily. Highly readable text, supplemented by 594 illustrations of costumes of the ancient Mediterranean peoples, Greece and Rome, the Teutonic prehistoric period; costumes of the Middle Ages, Renaissance, Baroque, 18th and 19th centuries. Clear, measured patterns are provided for many clothing articles. Approach is practical throughout. Enlarged by Emma von Sichart. 464pp. 21030-8 Paperbound $3.50

ORIENTAL RUGS, ANTIQUE AND MODERN, Walter A. Hawley. A complete and authoritative treatise on the Oriental rug—where they are made, by whom and how, designs and symbols, characteristics in detail of the six major groups, how to distinguish them and how to buy them. Detailed technical data is provided on periods, weaves, warps, wefts, textures, sides, ends and knots, although no technical background is required for an understanding. 11 color plates, 80 halftones, 4 maps. vi + 320pp. 6⅛ x 9⅛. 22366-3 Paperbound $5.00

TEN BOOKS ON ARCHITECTURE, Vitruvius. By any standards the most important book on architecture ever written. Early Roman discussion of aesthetics of building, construction methods, orders, sites, and every other aspect of architecture has inspired, instructed architecture for about 2,000 years. Stands behind Palladio, Michelangelo, Bramante, Wren, countless others. Definitive Morris H. Morgan translation. 68 illustrations. xii + 331pp. 20645-9 Paperbound $3.50

THE FOUR BOOKS OF ARCHITECTURE, Andrea Palladio. Translated into every major Western European language in the two centuries following its publication in 1570, this has been one of the most influential books in the history of architecture. Complete reprint of the 1738 Isaac Ware edition. New introduction by Adolf Placzek, Columbia Univ. 216 plates. xxii + 110pp. of text. 9½ x 12¾.
21308-0 Clothbound $10.00

STICKS AND STONES: A STUDY OF AMERICAN ARCHITECTURE AND CIVILIZATION, Lewis Mumford.One of the great classics of American cultural history. American architecture from the medieval-inspired earliest forms to the early 20th century; evolution of structure and style, and reciprocal influences on environment. 21 photographic illustrations. 238pp. 20202-X Paperbound $2.00

THE AMERICAN BUILDER'S COMPANION, Asher Benjamin. The most widely used early 19th century architectural style and source book, for colonial up into Greek Revival periods. Extensive development of geometry of carpentering, construction of sashes, frames, doors, stairs; plans and elevations of domestic and other buildings. Hundreds of thousands of houses were built according to this book, now invaluable to historians, architects, restorers, etc. 1827 edition. 59 plates. 114pp. 7⅞ x 10¾.
22236-5 Paperbound $3.50

DUTCH HOUSES IN THE HUDSON VALLEY BEFORE 1776, Helen Wilkinson Reynolds. The standard survey of the Dutch colonial house and outbuildings, with constructional features, decoration, and local history associated with individual homesteads. Introduction by Franklin D. Roosevelt. Map. 150 illustrations. 469pp. 6⅝ x 9¼. 21469-9 Paperbound $4.00

THE ARCHITECTURE OF COUNTRY HOUSES, Andrew J. Downing. Together with Vaux's *Villas and Cottages* this is the basic book for Hudson River Gothic architecture of the middle Victorian period. Full, sound discussions of general aspects of housing, architecture, style, decoration, furnishing, together with scores of detailed house plans, illustrations of specific buildings, accompanied by full text. Perhaps the most influential single American architectural book. 1850 edition. Introduction by J. Stewart Johnson. 321 figures, 34 architectural designs. xvi + 560pp.
22003-6 Paperbound $4.00

LOST EXAMPLES OF COLONIAL ARCHITECTURE, John Mead Howells. Full-page photographs of buildings that have disappeared or been so altered as to be denatured, including many designed by major early American architects. 245 plates. xvii + 248pp. 7⅞ x 10¾. 21143-6 Paperbound $3.50

DOMESTIC ARCHITECTURE OF THE AMERICAN COLONIES AND OF THE EARLY REPUBLIC, Fiske Kimball. Foremost architect and restorer of Williamsburg and Monticello covers nearly 200 homes between 1620-1825. Architectural details, construction, style features, special fixtures, floor plans, etc. Generally considered finest work in its area. 219 illustrations of houses, doorways, windows, capital mantels. xx + 314pp. 7⅞ x 10¾. 21743-4 Paperbound $4.00

EARLY AMERICAN ROOMS: 1650-1858, edited by Russell Hawes Kettell. Tour of 12 rooms, each representative of a different era in American history and each furnished, decorated, designed and occupied in the style of the era. 72 plans and elevations, 8-page color section, etc., show fabrics, wall papers, arrangements, etc. Full descriptive text. xvii + 200pp. of text. 8⅜ x 11¼.
21633-0 Paperbound $5.00

THE FITZWILLIAM VIRGINAL BOOK, edited by J. Fuller Maitland and W. B. Squire. Full modern printing of famous early 17th-century ms. volume of 300 works by Morley, Byrd, Bull, Gibbons, etc. For piano or other modern keyboard instrument; easy to read format. xxxvi + 938pp. 8⅜ x 11.
21068-5, 21069-3 Two volumes, Paperbound $10.00

KEYBOARD MUSIC, Johann Sebastian Bach. Bach Gesellschaft edition. A rich selection of Bach's masterpieces for the harpsichord: the six English Suites, six French Suites, the six Partitas (Clavierübung part I), the Goldberg Variations (Clavierübung part IV), the fifteen Two-Part Inventions and the fifteen Three-Part Sinfonias. Clearly reproduced on large sheets with ample margins; eminently playable. vi + 312pp. 8⅛ x 11. 22360-4 Paperbound $5.00

THE MUSIC OF BACH: AN INTRODUCTION, Charles Sanford Terry. A fine, nontechnical introduction to Bach's music, both instrumental and vocal. Covers organ music, chamber music, passion music, other types. Analyzes themes, developments, innovations. x + 114pp. 21075-8 Paperbound $1.25

BEETHOVEN AND HIS NINE SYMPHONIES, Sir George Grove. Noted British musicologist provides best history, analysis, commentary on symphonies. Very thorough, rigorously accurate; necessary to both advanced student and amateur music lover. 436 musical passages. vii + 407 pp. 20334-4 Paperbound $2.75

JOHANN SEBASTIAN BACH, Philipp Spitta. One of the great classics of musicology, this definitive analysis of Bach's music (and life) has never been surpassed. Lucid, nontechnical analyses of hundreds of pieces (30 pages devoted to St. Matthew Passion, 26 to B Minor Mass). Also includes major analysis of 18th-century music. 450 musical examples. 40-page musical supplement. Total of xx + 1799pp.
(EUK) 22278-0, 22279-9 Two volumes, Clothbound $17.50

MOZART AND HIS PIANO CONCERTOS, Cuthbert Girdlestone. The only full-length study of an important area of Mozart's creativity. Provides detailed analyses of all 23 concertos, traces inspirational sources. 417 musical examples. Second edition. 509pp. (USO) 21271-8 Paperbound $3.50

THE PERFECT WAGNERITE: A COMMENTARY ON THE NIBLUNG'S RING, George Bernard Shaw. Brilliant and still relevant criticism in remarkable essays on Wagner's Ring cycle, Shaw's ideas on political and social ideology behind the plots, role of Leitmotifs, vocal requisites, etc. Prefaces. xxi + 136pp.
21707-8 Paperbound $1.50

DON GIOVANNI, W. A. Mozart. Complete libretto, modern English translation; biographies of composer and librettist; accounts of early performances and critical reaction. Lavishly illustrated. All the material you need to understand and appreciate this great work. Dover Opera Guide and Libretto Series; translated and introduced by Ellen Bleiler. 92 illustrations. 209pp.
21134-7 Paperbound $2.00

HIGH FIDELITY SYSTEMS: A LAYMAN'S GUIDE, Roy F. Allison. All the basic information you need for setting up your own audio system: high fidelity and stereo record players, tape records, F.M. Connections, adjusting tone arm, cartridge, checking needle alignment, positioning speakers, phasing speakers, adjusting hums, trouble-shooting, maintenance, and similar topics. Enlarged 1965 edition. More than 50 charts, diagrams, photos. iv + 91pp. 21514-8 Paperbound $1.25

REPRODUCTION OF SOUND, Edgar Villchur. Thorough coverage for laymen of high fidelity systems, reproducing systems in general, needles, amplifiers, preamps, loudspeakers, feedback, explaining physical background. "A rare talent for making technicalities vividly comprehensible," R. Darrell, *High Fidelity*. 69 figures. iv + 92pp. 21515-6 Paperbound $1.25

HEAR ME TALKIN' TO YA: THE STORY OF JAZZ AS TOLD BY THE MEN WHO MADE IT, Nat Shapiro and Nat Hentoff. Louis Armstrong, Fats Waller, Jo Jones, Clarence Williams, Billy Holiday, Duke Ellington, Jelly Roll Morton and dozens of other jazz greats tell how it was in Chicago's South Side, New Orleans, depression Harlem and the modern West Coast as jazz was born and grew. xvi + 429pp.
21726-4 Paperbound $2.50

FABLES OF AESOP, translated by Sir Roger L'Estrange. A reproduction of the very rare 1931 Paris edition; a selection of the most interesting fables, together with 50 imaginative drawings by Alexander Calder. v + 128pp. 6½x9¼.
21780-9 Paperbound $1.50

AGAINST THE GRAIN (A REBOURS), Joris K. Huysmans. Filled with weird images, evidences of a bizarre imagination, exotic experiments with hallucinatory drugs, rich tastes and smells and the diversions of its sybarite hero Duc Jean des Esseintes, this classic novel pushed 19th-century literary decadence to its limits. Full unabridged edition. Do not confuse this with abridged editions generally sold. Introduction by Havelock Ellis. xlix + 206pp. 22190-3 Paperbound $2.00

VARIORUM SHAKESPEARE: HAMLET. Edited by Horace H. Furness; a landmark of American scholarship. Exhaustive footnotes and appendices treat all doubtful words and phrases, as well as suggested critical emendations throughout the play's history. First volume contains editor's own text, collated with all Quartos and Folios. Second volume contains full first Quarto, translations of Shakespeare's sources (Belleforest, and Saxo Grammaticus), Der Bestrafte Brudermord, and many essays on critical and historical points of interest by major authorities of past and present. Includes details of staging and costuming over the years. By far the best edition available for serious students of Shakespeare. Total of xx + 905pp.
21004-9, 21005-7, 2 volumes, Paperbound $7.00

A LIFE OF WILLIAM SHAKESPEARE, Sir Sidney Lee. This is the standard life of Shakespeare, summarizing everything known about Shakespeare and his plays. Incredibly rich in material, broad in coverage, clear and judicious, it has served thousands as the best introduction to Shakespeare. 1931 edition. 9 plates. xxix + 792pp. (USO) 21967-4 Paperbound $3.75

MASTERS OF THE DRAMA, John Gassner. Most comprehensive history of the drama in print, covering every tradition from Greeks to modern Europe and America, including India, Far East, etc. Covers more than 800 dramatists, 2000 plays, with biographical material, plot summaries, theatre history, criticism, etc. "Best of its kind in English," *New Republic.* 77 illustrations. xxii + 890pp.
20100-7 Clothbound $8.50

THE EVOLUTION OF THE ENGLISH LANGUAGE, George McKnight. The growth of English, from the 14th century to the present. Unusual, non-technical account presents basic information in very interesting form: sound shifts, change in grammar and syntax, vocabulary growth, similar topics. Abundantly illustrated with quotations. Formerly *Modern English in the Making.* xii + 590pp.
21932-1 Paperbound $3.50

AN ETYMOLOGICAL DICTIONARY OF MODERN ENGLISH, Ernest Weekley. Fullest, richest work of its sort, by foremost British lexicographer. Detailed word histories, including many colloquial and archaic words; extensive quotations. Do not confuse this with the Concise Etymological Dictionary, which is much abridged. Total of xxvii + 830pp. 6½ x 9¼.
21873-2, 21874-0 Two volumes, Paperbound $6.00

FLATLAND: A ROMANCE OF MANY DIMENSIONS, E. A. Abbott. Classic of science-fiction explores ramifications of life in a two-dimensional world, and what happens when a three-dimensional being intrudes. Amusing reading, but also useful as introduction to thought about hyperspace. Introduction by Banesh Hoffmann. 16 illustrations. xx + 103pp. 20001-9 Paperbound $1.00

Poems of Anne Bradstreet, edited with an introduction by Robert Hutchinson. A new selection of poems by America's first poet and perhaps the first significant woman poet in the English language. 48 poems display her development in works of considerable variety—love poems, domestic poems, religious meditations, formal elegies, "quaternions," etc. Notes, bibliography. viii + 222pp.
22160-1 Paperbound $2.00

Three Gothic Novels: The Castle of Otranto by Horace Walpole; Vathek by William Beckford; The Vampyre by John Polidori, with Fragment of a Novel by Lord Byron, edited by E. F. Bleiler. The first Gothic novel, by Walpole; the finest Oriental tale in English, by Beckford; powerful Romantic supernatural story in versions by Polidori and Byron. All extremely important in history of literature; all still exciting, packed with supernatural thrills, ghosts, haunted castles, magic, etc. xl + 291pp.
21232-7 Paperbound $2.50

The Best Tales of Hoffmann, E. T. A. Hoffmann. 10 of Hoffmann's most important stories, in modern re-editings of standard translations: Nutcracker and the King of Mice, Signor Formica, Automata, The Sandman, Rath Krespel, The Golden Flowerpot, Master Martin the Cooper, The Mines of Falun, The King's Betrothed, A New Year's Eve Adventure. 7 illustrations by Hoffmann. Edited by E. F. Bleiler. xxxix + 419pp. 21793-0 Paperbound $3.00

Ghost and Horror Stories of Ambrose Bierce, Ambrose Bierce. 23 strikingly modern stories of the horrors latent in the human mind: The Eyes of the Panther, The Damned Thing, An Occurrence at Owl Creek Bridge, An Inhabitant of Carcosa, etc., plus the dream-essay, Visions of the Night. Edited by E. F. Bleiler. xxii + 199pp. 20767-6 Paperbound $1.50

Best Ghost Stories of J. S. LeFanu, J. Sheridan LeFanu. Finest stories by Victorian master often considered greatest supernatural writer of all. Carmilla, Green Tea, The Haunted Baronet, The Familiar, and 12 others. Most never before available in the U. S. A. Edited by E. F. Bleiler. 8 illustrations from Victorian publications. xvii + 467pp. 20415-4 Paperbound $3.00

Mathematical Foundations of Information Theory, A. I. Khinchin. Comprehensive introduction to work of Shannon, McMillan, Feinstein and Khinchin, placing these investigations on a rigorous mathematical basis. Covers entropy concept in probability theory, uniqueness theorem, Shannon's inequality, ergodic sources, the E property, martingale concept, noise, Feinstein's fundamental lemma, Shanon's first and second theorems. Translated by R. A. Silverman and M. D. Friedman. iii + 120pp. 60434-9 Paperbound $1.75

Seven Science Fiction Novels, H. G. Wells. The standard collection of the great novels. Complete, unabridged. *First Men in the Moon, Island of Dr. Moreau, War of the Worlds, Food of the Gods, Invisible Man, Time Machine, In the Days of the Comet.* Not only science fiction fans, but every educated person owes it to himself to read these novels. 1015pp. 20264-X Clothbound $5.00

LAST AND FIRST MEN AND STAR MAKER, TWO SCIENCE FICTION NOVELS, Olaf Stapledon. Greatest future histories in science fiction. In the first, human intelligence is the "hero," through strange paths of evolution, interplanetary invasions, incredible technologies, near extinctions and reemergences. Star Maker describes the quest of a band of star rovers for intelligence itself, through time and space: weird inhuman civilizations, crustacean minds, symbiotic worlds, etc. Complete, unabridged. v + 438pp. 21962-3 Paperbound $2.50

THREE PROPHETIC NOVELS, H. G. WELLS. Stages of a consistently planned future for mankind. *When the Sleeper Wakes,* and *A Story of the Days to Come,* anticipate *Brave New World* and *1984,* in the 21st Century; *The Time Machine,* only complete version in print, shows farther future and the end of mankind. All show Wells's greatest gifts as storyteller and novelist. Edited by E. F. Bleiler. x + 335pp. (USO) 20605-X Paperbound $2.50

THE DEVIL'S DICTIONARY, Ambrose Bierce. America's own Oscar Wilde—Ambrose Bierce—offers his barbed iconoclastic wisdom in over 1,000 definitions hailed by H. L. Mencken as "some of the most gorgeous witticisms in the English language." 145pp. 20487-1 Paperbound $1.25

MAX AND MORITZ, Wilhelm Busch. Great children's classic, father of comic strip, of two bad boys, Max and Moritz. Also Ker and Plunk (Plisch und Plumm), Cat and Mouse, Deceitful Henry, Ice-Peter, The Boy and the Pipe, and five other pieces. Original German, with English translation. Edited by H. Arthur Klein; translations by various hands and H. Arthur Klein. vi + 216pp.
20181-3 Paperbound $2.00

PIGS IS PIGS AND OTHER FAVORITES, Ellis Parker Butler. The title story is one of the best humor short stories, as Mike Flannery obfuscates biology and English. Also included, That Pup of Murchison's, The Great American Pie Company, and Perkins of Portland. 14 illustrations. v + 109pp. 21532-6 Paperbound $1.25

THE PETERKIN PAPERS, Lucretia P. Hale. It takes genius to be as stupidly mad as the Peterkins, as they decide to become wise, celebrate the "Fourth," keep a cow, and otherwise strain the resources of the Lady from Philadelphia. Basic book of American humor. 153 illustrations. 219pp. 20794-3 Paperbound $1.50

PERRAULT'S FAIRY TALES, translated by A. E. Johnson and S. R. Littlewood, with 34 full-page illustrations by Gustave Doré. All the original Perrault stories—Cinderella, Sleeping Beauty, Bluebeard, Little Red Riding Hood, Puss in Boots, Tom Thumb, etc.—with their witty verse morals and the magnificent illustrations of Doré. One of the five or six great books of European fairy tales. viii + 117pp. 8⅛ x 11. 22311-6 Paperbound $2.00

OLD HUNGARIAN FAIRY TALES, Baroness Orczy. Favorites translated and adapted by author of the *Scarlet Pimpernel.* Eight fairy tales include "The Suitors of Princess Fire-Fly," "The Twin Hunchbacks," "Mr. Cuttlefish's Love Story," and "The Enchanted Cat." This little volume of magic and adventure will captivate children as it has for generations. 90 drawings by Montagu Barstow. 96pp.
(USO) 22293-4 Paperbound $1.95

THE RED FAIRY BOOK, Andrew Lang. Lang's color fairy books have long been children's favorites. This volume includes Rapunzel, Jack and the Bean-stalk and 35 other stories, familiar and unfamiliar. 4 plates, 93 illustrations x + 367pp.
21673-X Paperbound $2.50

THE BLUE FAIRY BOOK, Andrew Lang. Lang's tales come from all countries and all times. Here are 37 tales from Grimm, the Arabian Nights, Greek Mythology, and other fascinating sources. 8 plates, 130 illustrations. xi + 390pp.
21437-0 Paperbound $2.50

HOUSEHOLD STORIES BY THE BROTHERS GRIMM. Classic English-language edition of the well-known tales — Rumpelstiltskin, Snow White, Hansel and Gretel, The Twelve Brothers, Faithful John, Rapunzel, Tom Thumb (52 stories in all). Translated into simple, straightforward English by Lucy Crane. Ornamented with headpieces, vignettes, elaborate decorative initials and a dozen full-page illustrations by Walter Crane. x + 269pp. 21080-4 Paperbound $2.50

THE MERRY ADVENTURES OF ROBIN HOOD, Howard Pyle. The finest modern versions of the traditional ballads and tales about the great English outlaw. Howard Pyle's complete prose version, with every word, every illustration of the first edition. Do not confuse this facsimile of the original (1883) with modern editions that change text or illustrations. 23 plates plus many page decorations. xxii + 296pp.
22043-5 Paperbound $2.50

THE STORY OF KING ARTHUR AND HIS KNIGHTS, Howard Pyle. The finest children's version of the life of King Arthur; brilliantly retold by Pyle, with 48 of his most imaginative illustrations. xviii + 313pp. $6\frac{1}{8}$ x $9\frac{1}{4}$.
21445-1 Paperbound $2.50

THE WONDERFUL WIZARD OF OZ, L. Frank Baum. America's finest children's book in facsimile of first edition with all Denslow illustrations in full color. The edition a child should have. Introduction by Martin Gardner. 23 color plates, scores of drawings. iv + 267pp. 20691-2 Paperbound $2.50

THE MARVELOUS LAND OF OZ, L. Frank Baum. The second Oz book, every bit as imaginative as the Wizard. The hero is a boy named Tip, but the Scarecrow and the Tin Woodman are back, as is the Oz magic. 16 color plates, 120 drawings by John R. Neill. 287pp. 20692-0 Paperbound $2.50

THE MAGICAL MONARCH OF MO, L. Frank Baum. Remarkable adventures in a land even stranger than Oz. The best of Baum's books not in the Oz series. 15 color plates and dozens of drawings by Frank Verbeck. xviii + 237pp.
21892-9 Paperbound $2.25

THE BAD CHILD'S BOOK OF BEASTS, MORE BEASTS FOR WORSE CHILDREN, A MORAL ALPHABET, Hilaire Belloc. Three complete humor classics in one volume. Be kind to the frog, and do not call him names . . . and 28 other whimsical animals. Familiar favorites and some not so well known. Illustrated by Basil Blackwell. 156pp. (USO) 20749-8 Paperbound $1.50

EAST O' THE SUN AND WEST O' THE MOON, George W. Dasent. Considered the best of all translations of these Norwegian folk tales, this collection has been enjoyed by generations of children (and folklorists too). Includes True and Untrue, Why the Sea is Salt, East O' the Sun and West O' the Moon, Why the Bear is Stumpy-Tailed, Boots and the Troll, The Cock and the Hen, Rich Peter the Pedlar, and 52 more. The only edition with all 59 tales. 77 illustrations by Erik Werenskiold and Theodor Kittelsen. xv + 418pp. 22521-6 Paperbound $3.50

GOOPS AND HOW TO BE THEM, Gelett Burgess. Classic of tongue-in-cheek humor, masquerading as etiquette book. 87 verses, twice as many cartoons, show mischievous Goops as they demonstrate to children virtues of table manners, neatness, courtesy, etc. Favorite for generations. viii + 88pp. 6½ x 9¼.
22233-0 Paperbound $1.25

ALICE'S ADVENTURES UNDER GROUND, Lewis Carroll. The first version, quite different from the final *Alice in Wonderland,* printed out by Carroll himself with his own illustrations. Complete facsimile of the "million dollar" manuscript Carroll gave to Alice Liddell in 1864. Introduction by Martin Gardner. viii + 96pp. Title and dedication pages in color. 21482-6 Paperbound $1.25

THE BROWNIES, THEIR BOOK, Palmer Cox. Small as mice, cunning as foxes, exuberant and full of mischief, the Brownies go to the zoo, toy shop, seashore, circus, etc., in 24 verse adventures and 266 illustrations. Long a favorite, since their first appearance in St. Nicholas Magazine. xi + 144pp. 6⅝ x 9¼.
21265-3 Paperbound $1.75

SONGS OF CHILDHOOD, Walter De La Mare. Published (under the pseudonym Walter Ramal) when De La Mare was only 29, this charming collection has long been a favorite children's book. A facsimile of the first edition in paper, the 47 poems capture the simplicity of the nursery rhyme and the ballad, including such lyrics as I Met Eve, Tartary, The Silver Penny. vii + 106pp. 21972-0 Paperbound $1.25

THE COMPLETE NONSENSE OF EDWARD LEAR, Edward Lear. The finest 19th-century humorist-cartoonist in full: all nonsense limericks, zany alphabets, Owl and Pussycat, songs, nonsense botany, and more than 500 illustrations by Lear himself. Edited by Holbrook Jackson. xxix + 287pp. (USO) 20167-8 Paperbound $2.00

BILLY WHISKERS: THE AUTOBIOGRAPHY OF A GOAT, Frances Trego Montgomery. A favorite of children since the early 20th century, here are the escapades of that rambunctious, irresistible and mischievous goat—Billy Whiskers. Much in the spirit of *Peck's Bad Boy,* this is a book that children never tire of reading or hearing. All the original familiar illustrations by W. H. Fry are included: 6 color plates, 18 black and white drawings. 159pp. 22345-0 Paperbound $2.00

MOTHER GOOSE MELODIES. Faithful republication of the fabulously rare Munroe and Francis "copyright 1833" Boston edition—the most important Mother Goose collection, usually referred to as the "original." Familiar rhymes plus many rare ones, with wonderful old woodcut illustrations. Edited by E. F. Bleiler. 128pp. 4½ x 6⅜. 22577-1 Paperbound $1.25

TWO LITTLE SAVAGES; BEING THE ADVENTURES OF TWO BOYS WHO LIVED AS INDIANS AND WHAT THEY LEARNED, Ernest Thompson Seton. Great classic of nature and boyhood provides a vast range of woodlore in most palatable form, a genuinely entertaining story. Two farm boys build a teepee in woods and live in it for a month, working out Indian solutions to living problems, star lore, birds and animals, plants, etc. 293 illustrations. vii + 286pp.
20985-7 Paperbound $2.50

PETER PIPER'S PRACTICAL PRINCIPLES OF PLAIN & PERFECT PRONUNCIATION. Alliterative jingles and tongue-twisters of surprising charm, that made their first appearance in America about 1830. Republished in full with the spirited woodcut illustrations from this earliest American edition. 32pp. 4½ x 6⅜.
22560-7 Paperbound $1.00

SCIENCE EXPERIMENTS AND AMUSEMENTS FOR CHILDREN, Charles Vivian. 73 easy experiments, requiring only materials found at home or easily available, such as candles, coins, steel wool, etc.; illustrate basic phenomena like vacuum, simple chemical reaction, etc. All safe. Modern, well-planned. Formerly *Science Games for Children*. 102 photos, numerous drawings. 96pp. 6⅛ x 9¼.
21856-2 Paperbound $1.25

AN INTRODUCTION TO CHESS MOVES AND TACTICS SIMPLY EXPLAINED, Leonard Barden. Informal intermediate introduction, quite strong in explaining reasons for moves. Covers basic material, tactics, important openings, traps, positional play in middle game, end game. Attempts to isolate patterns and recurrent configurations. Formerly *Chess*. 58 figures. 102pp. (USO) 21210-6 Paperbound $1.25

LASKER'S MANUAL OF CHESS, Dr. Emanuel Lasker. Lasker was not only one of the five great World Champions, he was also one of the ablest expositors, theorists, and analysts. In many ways, his Manual, permeated with his philosophy of battle, filled with keen insights, is one of the greatest works ever written on chess. Filled with analyzed games by the great players. A single-volume library that will profit almost any chess player, beginner or master. 308 diagrams. xli x 349pp.
20640-8 Paperbound $2.75

THE MASTER BOOK OF MATHEMATICAL RECREATIONS, Fred Schuh. In opinion of many the finest work ever prepared on mathematical puzzles, stunts, recreations; exhaustively thorough explanations of mathematics involved, analysis of effects, citation of puzzles and games. Mathematics involved is elementary. Translated by F. Göbel. 194 figures. xxiv + 430pp. 22134-2 Paperbound $3.00

MATHEMATICS, MAGIC AND MYSTERY, Martin Gardner. Puzzle editor for Scientific American explains mathematics behind various mystifying tricks: card tricks, stage "mind reading," coin and match tricks, counting out games, geometric dissections, etc. Probability sets, theory of numbers clearly explained. Also provides more than 400 tricks, guaranteed to work, that you can do. 135 illustrations. xii + 176pp.
20338-2 Paperbound $1.50

MATHEMATICAL PUZZLES FOR BEGINNERS AND ENTHUSIASTS, Geoffrey Mott-Smith. 189 puzzles from easy to difficult—involving arithmetic, logic, algebra, properties of digits, probability, etc.—for enjoyment and mental stimulus. Explanation of mathematical principles behind the puzzles. 135 illustrations. viii + 248pp.
20198-8 Paperbound $1.75

PAPER FOLDING FOR BEGINNERS, William D. Murray and Francis J. Rigney. Easiest book on the market, clearest instructions on making interesting, beautiful origami. Sail boats, cups, roosters, frogs that move legs, bonbon boxes, standing birds, etc. 40 projects; more than 275 diagrams and photographs. 94pp.
20713-7 Paperbound $1.00

TRICKS AND GAMES ON THE POOL TABLE, Fred Herrmann. 79 tricks and games—some solitaires, some for two or more players, some competitive games—to entertain you between formal games. Mystifying shots and throws, unusual caroms, tricks involving such props as cork, coins, a hat, etc. Formerly *Fun on the Pool Table*. 77 figures. 95pp.
21814-7 Paperbound $1.00

HAND SHADOWS TO BE THROWN UPON THE WALL: A SERIES OF NOVEL AND AMUSING FIGURES FORMED BY THE HAND, Henry Bursill. Delightful picturebook from great-grandfather's day shows how to make 18 different hand shadows: a bird that flies, duck that quacks, dog that wags tail, camel, goose, deer, boy, turtle, etc. Only book of its sort. vi + 33pp. [illegible]
21779-5 Paperbound $1.00

WHITTLING AND WOODCARVING, E. J. Tangerman. 18th printing of best book on market. "If you can cut a potato you can carve" toys and puzzles, chains, chessmen, caricatures, masks, frames, woodcut blocks, surface patterns, much more. Information on tools, woods, techniques. Also goes into serious wood sculpture from Middle Ages to present, East and West. 464 photos, figures. x + 293pp.
20965-2 Paperbound $2.00

HISTORY OF PHILOSOPHY, Julián Marías. Possibly the clearest, most easily followed, best planned, most useful one-volume history of philosophy on the market; neither skimpy nor overfull. Full details on system of every major philosopher and dozens of less important thinkers from pre-Socratics up to Existentialism and later. Strong on many European figures usually omitted. Has gone through dozens of editions in Europe. 1966 edition, translated by Stanley Appelbaum and Clarence Strowbridge. xviii + 505pp.
21739-6 Paperbound $3.00

YOGA: A SCIENTIFIC EVALUATION, Kovoor T. Behanan. Scientific but non-technical study of physiological results of yoga exercises; done under auspices of Yale U. Relations to Indian thought, to psychoanalysis, etc. 16 photos. xxiii + 270pp.
20505-3 Paperbound $2.50

Prices subject to change without notice.
Available at your book dealer or write for free catalogue to Dept. Adsi, Dover Publications, Inc., 180 Varick St., N.Y., N.Y. 10014. Dover publishes more than 150 books each year on science, elementary and advanced mathematics, biology, music, art, literary history, social sciences and other areas.